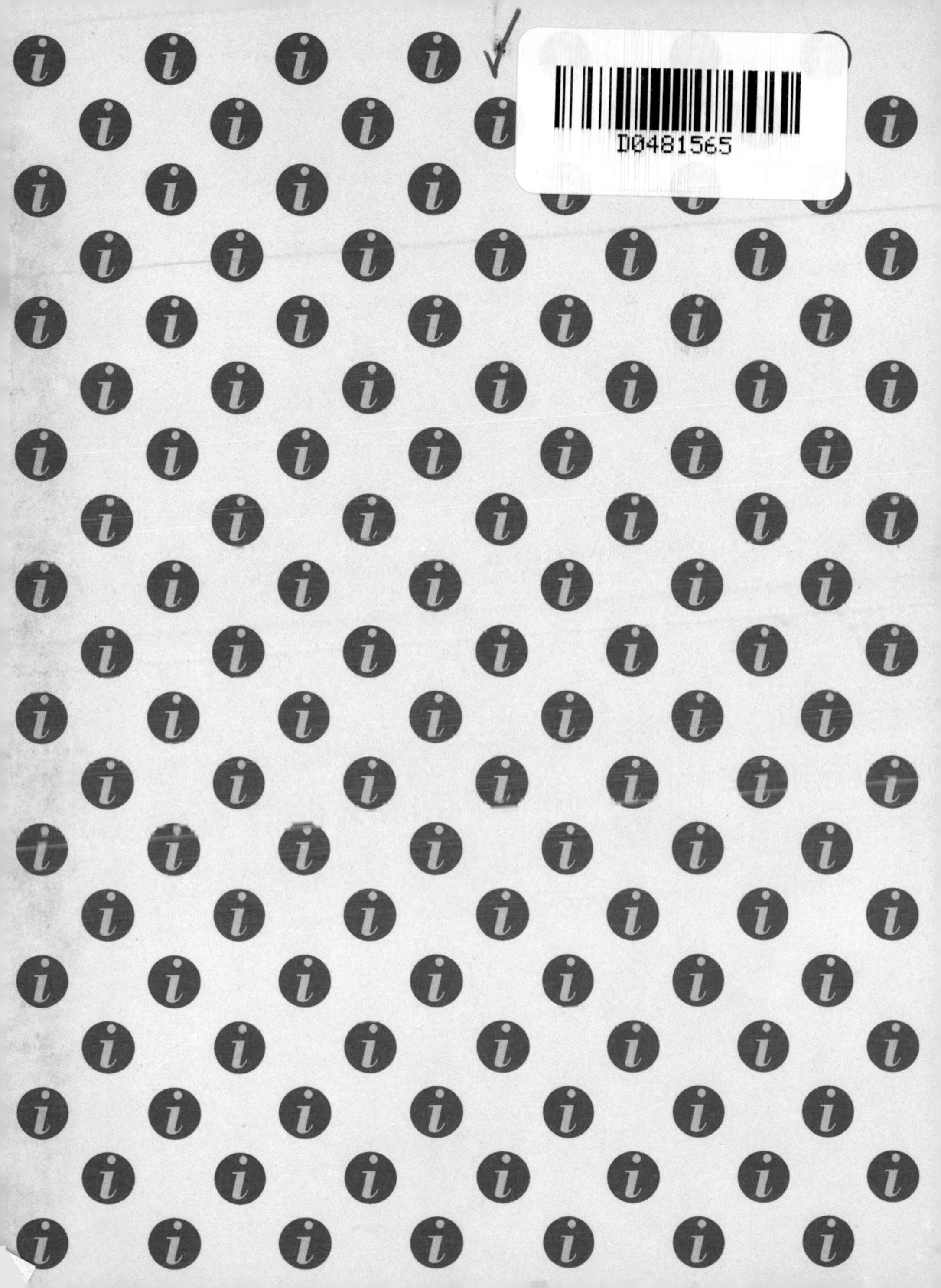
D0481565

IDENTIFYING

i

CRYSTALS

The new compact study guide and identifier

The new compact study guide and identifier

Peter Darling

CHARTWELL BOOKS, INC.

A QUINTET BOOK

Published by Chartwell Books
A Division of Book Sales, Inc.
114, Northfield Avenue
Edison, New Jersey 08837

This edition produced for sale in the U.S.A., its
territories and dependencies only.

ISBN 0-7858-0945-7

This book was designed and produced by
Quintet Publishing Limited
6 Blundell Street
London N7 9BH

Creative Director: Richard Dewing
Art Director: Silke Braun
Designer: James Lawrence
Project Editor: Clare Hubbard
Editor: Rosie Hankin

Typeset in Great Britain by
Central Southern Typesetters, Eastbourne
Manufactured in Singapore by Eray Scan Pte Ltd
Printed in Singapore by
Star Standard Industries (Pte) Ltd

The material used in this publication
previously appeared in *Crystal Identifier, Identifying Gems and Precious Stones*, and *Identifying Rocks and Minerals.*

Picture Credits
l = left; r = right; c = center; a = above; b = below
British Museum: 18 b.
British Museum (Natural History): 12 r, 14, 23 b, 41 a, 42 b, 47 a, 63 b.
C. M. Dixon, Canterbury: 18 a.
Geoscience Features Picture Library: 34 b, 39 b, 40 b, 51 a, 52 b, 60 a, 61 a, 64 b, 65 a, 65 b, 67 a, 67 b, 68 b, 69 b, 74 a, 74 b.
Geoscience Features Picture Library/A. Fisher: 22 b, 25 b, 26 a, 26 b, 27 b, 28 a, 34 a, 35 b, 37 a, 38 a, 39 a, 40 a, 49 b, 52 a, 56 a, 56 b, 57 a, 58 a, 59 a, 59 b, 60 b, 62 b, 64 a, 68 a, 73 a, 76 b.
Geoscience Features Picture Library/Dr B. Booth: 7, 8 l, 24 a, 24 b, 27 a, 29 a, 29 b, 30 a, 30 b, 31 a, 31 b, 33 a, 33 b, 35 a, 36 a, 36 b, 38 b, 41 b, 43 a, 43 b, 44 b, 45 a, 45 b, 46 b, 47 b, 48 a, 48 b, 49 a, 50 a, 53 a, 53 b, 54 a, 54 b, 55 b, 57 b, 61 b, 62 a, 66 a, 66 b, 69 a, 70 a, 70 b, 71 a, 71 b, 72 a, 72 b, 73 b, 75 a, 75 b, 76 a.
Geoscience Features Picture Library/W. Hughes: 19 a, 19 b.
Natural History Museum, London: 22 a, 23 a, 25 a, 28 b, 32 a, 37 b, 42 a, 46 a, 50 b, 51 b, 55 a, 63 a.
Peter Nixon (University of Leeds): 6, 8 c, 9, 12 l, 13, 20 a, 20 b, 32 b, 44 a, 58 b.

Contents

Introduction

The Crystal

Crystals have excited imagination and desire for thousands of years. Created by chance in unbelievably hostile environments, crystals have a form precise enough to delight scientists, colors variable enough to inspire artists, and a chemical make-up as unpredictable and fascinating as the weather.

It is no wonder that crystals have been regarded with awe by people since the dawn of time. Early civilizations imbued crystals with supernatural properties, ready to discharge their powers in the service of their master. It is also believed that they were among the first materials to be chipped and fashioned and then lashed onto sticks to form crude but effective weapons.

Crystals have been used for centuries to ward off evil and cure all manner of ailments. They appeared in the crowns of bishops and kings, in love tokens, and in amulets. On a practical level, they were a convenient way to store and show off wealth.

Today, we have a more complete and scientific understanding of the origin and nature of crystals. They can most simply be defined as naturally occurring inorganic minerals which have taken on a uniform shape within a specific structure. Their physical and chemical properties are defined by the elements in which they are formed. However, even now, many people believe that crystals can play a part in providing relief and hope to those in mental or physical distress. Just as homeopathy is used as a supplement or alternative to conventional medicine, so crystals maintain their place alongside the surgical scalpel and the bottle of pills.

One of the most common crystals is rock salt or halite. Besides being used in food preparation, it is the major source of the elements of soda, sodium, and hydrochloric acid, without which the chemical industry could not exist. There are salt deposits hundreds of feet below the surface of the waters in the Gulf of Mexico that have evolved into

below: *This eighteenth-century soapstone figure is Shoulao, the Chinese god of longevity, and illustrates one of the many associations between crystals and good luck charms.*

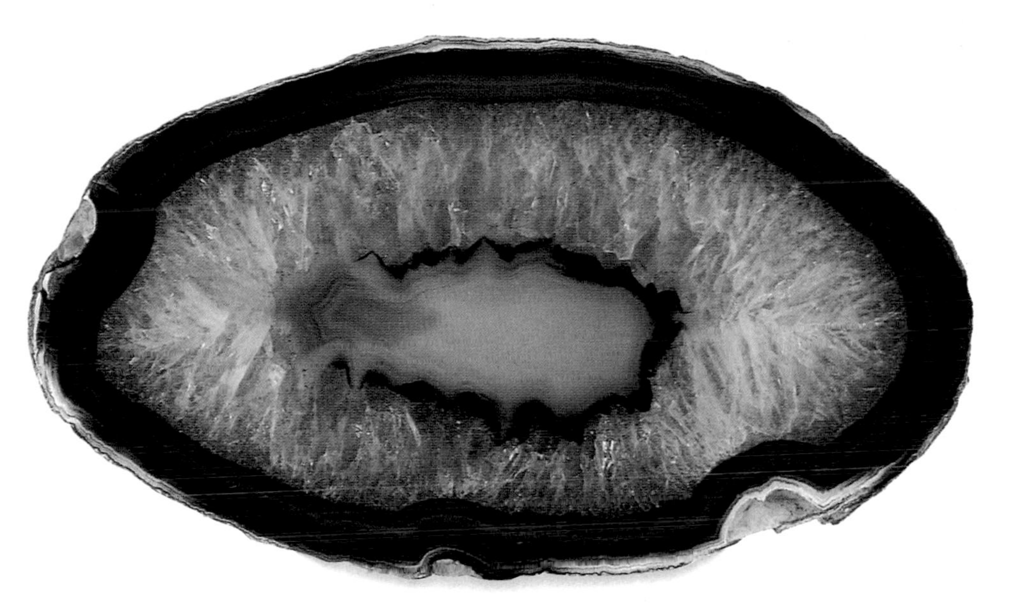

above: *This fine example of a Brazilian agate geode has been infilled with quartz and then, when that layer was completed, a central core of opal.*

massive dome structures. These deposits add the tangibility of value to the fancy of spiritualism, and are one of the world's most abundant sources of salt, sulfur, petroleum, and natural gas.

Crystal Formation

How is it that something as perfect, beautiful, and useful as crystal came from the Earth where it is usually found among unimpressive-looking rocks?

With a few exceptions, minerals are not created but grow from tiny nuclei within a solution into crystals. It is still unclear how the process starts off. It may be a split-second, chance orientation of atoms into a solid form that proves enough to form the nuclei and start the crystal growth process.

At this stage, crystals grow rather like pearls, which develop layer by layer from a grain of sand within an oyster. It is hard to believe that crystals form one atom at a time. The process is repeated thousands of times per second and, depending on the size of the crystal, they can take anything from a day through a year to be completed.

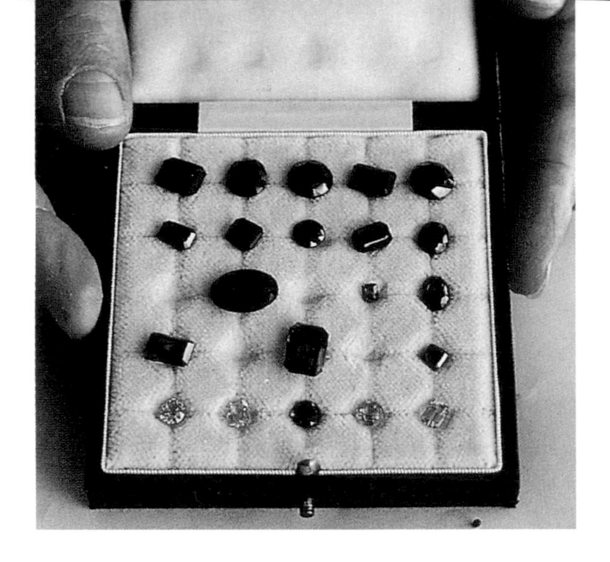

above: *A selection of Australian cut sapphires, the country's most profitable crystal.*

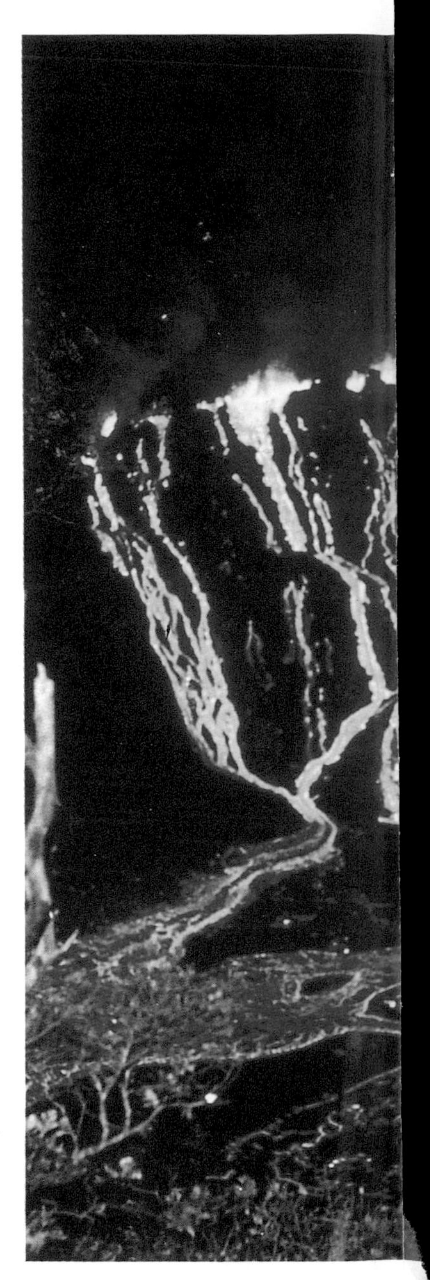

Crystal Solutions

Crystals are formed from gases, molten rock, or aqueous solutions which are created far below the Earth's surface. Sometimes they are re-formed from previously solid material which has been heated and pressurized until it liquifies, only resuming a different solid state when the pressure and/or heat source is removed. Such re-formed crystals may not resemble their original form due to the addition or subtraction of elements, or because of a change in the heat and/or pressure of the growth environment.

Crystals are often formed as a result of hydrothermal actions. A super-hot solution, heavily charged with chemical elements, is forced by high pressure through microcracks and veins. As the solution is displaced, its temperature and pressure are dissipated, and crystals will grow under these less "turbulent" conditions.

Crystals are also formed from solutions of elemental salts which become steadily more concentrated as they exceed their saturation level. This usually happens around inland lakes or seas after geological uplift. As natural evaporation occurs, solutions become more concentrated until saturation occurs, and crystallization follows.

Growth Rate

The size of crystals depends on the rate of growth. The slower they grow, the larger they will be. A slower rate of growth may be due to evaporation, a cooling off of the solution, or the reduction of pressure. Sometimes the results are spectacular. In Brazil a beryl crystal weighing 200 tons was found and in Siberia a milky quartz crystal of 13 tons was uncovered. Slow growth also results in a more regular-shaped crystal as the atoms have more time to assume an orderly arrangement.

The faces of a crystal do not necessarily grow at the same rate, which can result in elongated crystals. A faster accumulation of atoms in one direction, perhaps due to a strong electrical attraction, will be exaggerated as the atoms are being attracted to a small face-end rather than to a long side. Similarly, if a crystal grows from a solid base, as is so often the case with amethysts and halite, only half of the crystal is in a position to receive new unit-cells. It is astonishing these crystals are not more commonly malformed and irregular in shape.

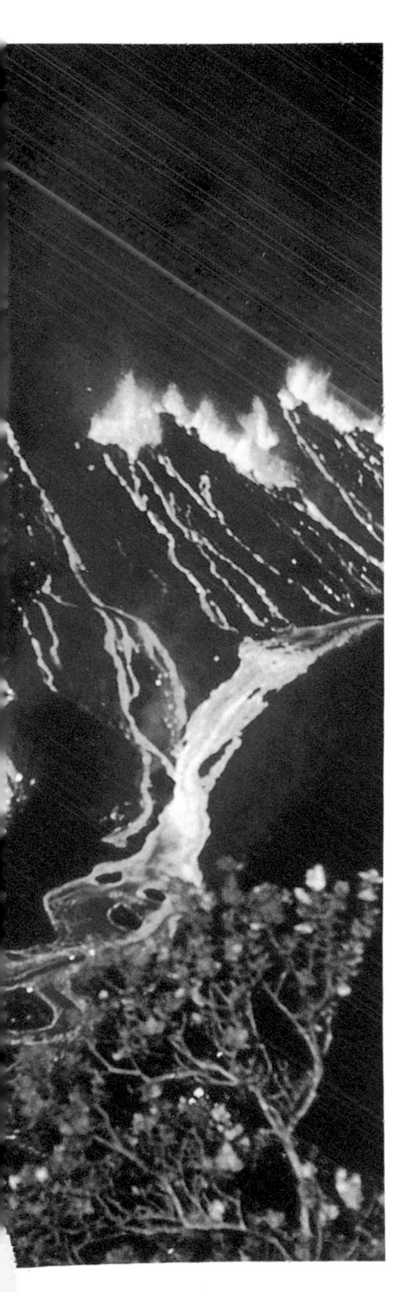

left: *Like most crystals, this eruption of lava at Kilauea Iki, Hawaii, owes its existence to activity originating deep within the Earth's crust.*

below: *The Oppenheimer diamond, held at the Smithsonian Institution, Washington, weighs 254 carats.*

Distinguishing Features of Crystals

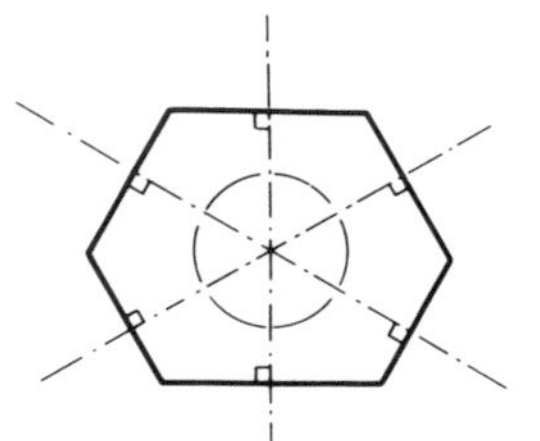

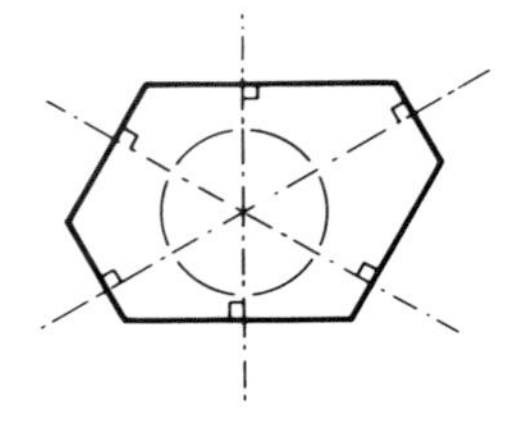

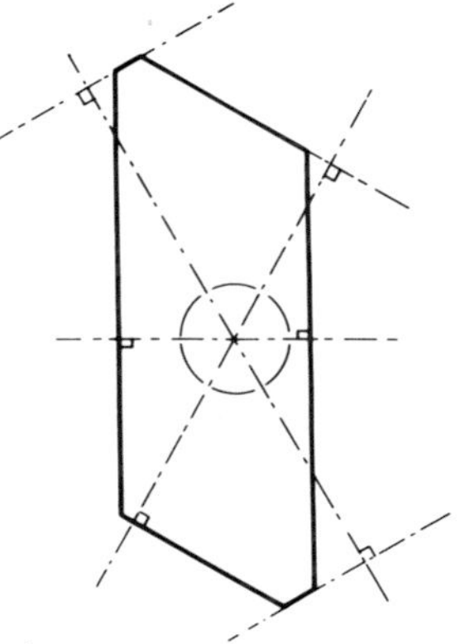

above: *Different sections through the main axes of quartz crystals. The crystal on the left has developed regularly while the other two have developed unequally but still comply with the law of constancy of interfacial angles.*

The regularity with which crystals form was first noted by Nicolaus Steno, the Danish scientist, in 1669. He found that at a specific temperature and pressure, all crystals of the same substance possess the same angle between corresponding faces. This is known as the law of constancy of interfacial angles. It was further clarified by Reinhard Bernhardi, in the early nineteenth century, who showed that the angle referred to was taken at right angles to the faces toward the central point of the crystal.

One of the truly fascinating and almost unbelievable aspects of crystals is that, despite their seemingly limitless shapes, they belong to one of only seven classifications of symmetry, categorized by their crystal axes, the angle at which the axes radiate from the crystal's central point of intersection, and the planes in which these axes lie. Even so, there are few "copy-book" examples. Many minerals occur in a combination of mineral forms, called polymorphs, of which calcite has 80. Most crystals favor one or two classic forms.

‑ystal Habit

number and orientation of faces that a crystal has function of the atomic structure of the elements environment in which the crystal solution ‑e environment, which is the balance between ‑re, pressure, and chemical saturation of the the solution, will have a marked effect on ‑rystals that eventually emerge.

Classification	Face Shape and Axis Orientation	Form
Isometric	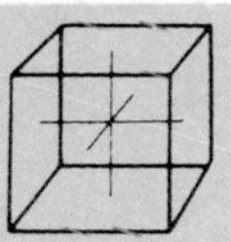	Three axes of equal length, at right angles to each other.
Tetragonal	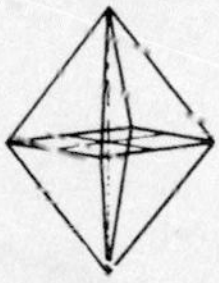	Three axes at right angles to each other: two on the same plane of equal length, the third perpendicular to them.
Hexagonal	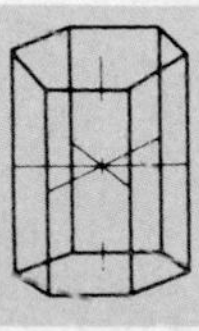	Three axes on one plane radiate out equally from a central point. The fourth axis is at right angles and is unequal in length. Six distinctive planes of symmetry parallel to the long axis.
Trigonal	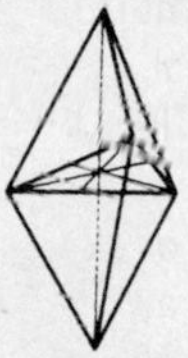	Three equal axes radiating from a single point in the same plane. A fourth axis is at right angles. Three distinct planes of symmetry parallel to the long axis.
Orthorhombic	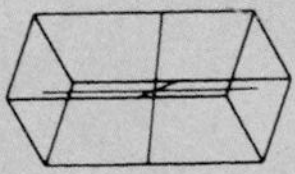	Three axes of unequal length set at right angles to one another.
Monoclinic	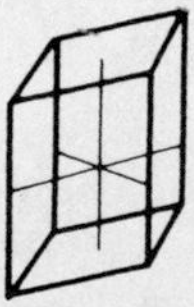	Prism with inclined top and bottom faces. Three axes of unequal length: two at right angles to each other, the third set at an incline to the plane of the others.
Triclinic	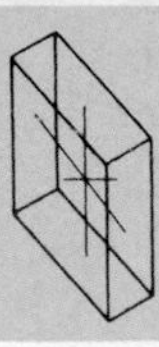	Three axes of unequal length set at three different angles to each other. Three pairs of faces.

Although in theory there is no limit to the possible shapes that a crystal will assume, generally crystals of a particular type will tend to grow in a regular fashion or "crystal habit." This term is used to describe the size and shape most frequently taken by the crystal. It is a result of a compromise between the way the solution would naturally crystallize out, given ideal conditions, and the environment in which the solution most often occurs.

Striations

These are ridges, furrows, or linear marks believed to be due to an oscillation of growth between two crystal orientations. This leads to less well-defined crystal edges. Rounded faces are characteristic of such behavior, and are especially common in tourmaline crystals.

above: *This fine example of feldspar variety adularia from Switzerland shows clearly the striations that may occur on the surface of a crystal.*

left: *Schists, such as this kyanite, are rocks which contain mineral deposits in parallel or subparallel veins.*

Chemical Impurities in Crystals

The presence of minute chemical impurities—which may possibly act as the initial nuclei—is often thought to cause a slowing up of crystal growth, giving a different crystal habit than that which normally occurs. Impurities can often manifest themselves as inclusions or striking, distinctive colors. They can be very beautiful and have often turned an undistinguished crystal into a unique object of admiration.

Color

Crystals with similar structures but different chemical compositions are often quite different in color and shape, leading to a marked variation in value, especially in gemstones. For example, the ruby and the sapphire are both members of the corundum group; rubies owe their color to traces of chrome and sapphires to minute quantities of iron and titanium. Because chrome is less common than iron, rubies are valued more highly.

Color is produced by reflected light. White light comprises electromagnetic pulses with seven different wavelengths, which produce the colors of the spectrum (red, orange, yellow, green, blue, indigo, and violet). These combined colors enter the crystal and are absorbed or partly absorbed by an element within. The metal elements chrome, cobalt, copper, iron, magnesium, nickel, titanium, and vanadium are common wavelength absorbers and therefore color creators.

If all wavelengths pass through the crystal, it appears colorless. If all wavelengths are absorbed, it appears black. If they are all absorbed to the same degree, it will appear dull or gray. Absorption and color can also be influenced by the distance light has to travel through a stone. This knowledge is useful to gem cutters who can manipulate the color of stones by leaving them thick (to enhance coloration) or hollowing them out (to lessen coloration).

Artificial light acts differently than natural light. For example, it enhances the beauty of emeralds and rubies but diminishes the appeal of sapphires. An extreme example is alexandrite, where the most obvious color change occurs. It appears green in natural light and red in artificial light.

below: *A crystal sample, put under the polarizing microscope, shows clearly the areas of different mineral types that appear as different color zones.*

Hardness

Hardness, an inherent and easily determined characteristic of a mineral, is measured using a resistance to scratching method, the Mohs' Scale. This is named for Friedrich Mohs (1773–1839), a Viennese mineralogist. He was frustrated by the imprecise terminology used to describe minerals, and set about to create a universal scale, hoping that it would help mineralogy to be considered a true science.

The scale runs from 1 (softest) to 10 (hardest) with each level being represented by a mineral. Each mineral in the series can scratch the mineral below it and can be scratched by the mineral above it. Traditionally a geologist would carry a sample of each of the minerals in the index to carry out field tests. Today, you can buy a set of hardness pencils which consist of splinters of the minerals set in a metal holder. When carrying out a hardness test, only test a sound surface and then only with a sharp point of the test mineral.

below: *Calcite from Iceland demonstrates the optical characteristic birefringence, in which an object viewed through the crystal is seen double.*

Mohs' Scale is empirical and does not bear any relationship to hardness in the strict scientific sense. This can be seen by consulting the Rosiwal's hardness table on the right-hand side of the table below, which is a scientific scale of hardness. This was devised by August Karl Rosiwal (1860–1923).

The hardness of a material depends on the atomic bonding of the crystal structure, since these bonds can vary depending on the crystallographic direction. Hardness variations are displayed y striated, laminated, or weathered crystals, among hers. The most celebrated example of hardness variation rs in the kyanite crystal which displays a hardness of 5 along its crystal axis but a value of 6 to 7 across stal axis.

Moh's Scale	Hardness	Comparison Mineral	Mineral Test	Rosiwal's Grinding Hardness
1	Soft	Talc	Powdered by fingernail	0.03
2	Soft	Gypsum	Scratched by fingernail	1.25
3	Medium hard	Calcite	Scratched by copper coin	4.5
4	Medium-hard	Fluorite	Easily scratched by pocket knife	5.0
5	Medium-hard	Apatite	Just scratched by pocket knife	6.5
6	Medium-hard	Orthoclase	Scratched by steel file	37
7	Hard	Quartz	Scratches glass window	120
8	Hard	Topaz	Easily scratches quartz	175
9	"Precious-stone" hard	Corundum	Easily scratches topaz	1,000
10	"Precious-stone" hard	Diamond	Cannot be scratched	140,000

Luster

This is concerned with a mineral's surface and the intensity with which that surface reflects light. A mirror would have perfect luster, described as "brilliant," while most materials, such as clay, are matte in appearance and are described as "dull."

The two main types of luster are metallic and nonmetallic. Fresh-cut metals and most sulfide materials have metallic luster, but there is a host of terms for those found in nonmetallic material. "Adamatine" is a brilliant luster found in such crystals as diamond and cassiterite. "Vitreous" is the most common form of luster, and can be described as glasslike, and is commonly found in quartz and beryl. A "resinous" luster is found in sulfur, while fibrous minerals such as gypsum and asbestos display a "silky" luster. Minerals with a "waxy" luster reflect only a small portion of the light hitting their surface, while those with an "earthy" or "dull" luster are normally aggregates or have rough weathered surfaces, such as montmorillonite and bauxite.

Cleavage

As explained earlier, a crystal is made up of atoms arranged in atomic units that join into a self-determining structure. The framework is a three-dimensional network of atomic bonds; these collectively form a crystal lattice. The bonds are not necessarily of equal strength. When subjected to mechanical stress, a crystal will tend to break cleanly along a lattice plane if the atomic bonds are strong in all but one direction. This crystal has perfect cleavage. When a crystal does not show a dominant direction of weak or strong bonds, it will not break cleanly, and is said to have poorly defined cleavage.

The presence of a cleavage plane is an inherent part of a crystal's composition. It is a characteristic often exploited when working with hard and precious stones to facilitate the process of shaping them.

Specific Gravity

The specific gravity (sometimes referred to as the relative density) of a crystal is defined as the weight per unit volume; in other words, the ratio of the weight of an object to that of an equal volume of distilled water. So a crystal with a specific gravity of 4 is four times as heavy as the same volume of distilled water.

The specific gravity of a crystal can be measured in a number of ways. To measure specific gravity using a

below: *Diagram of a hydrostatic balance.*

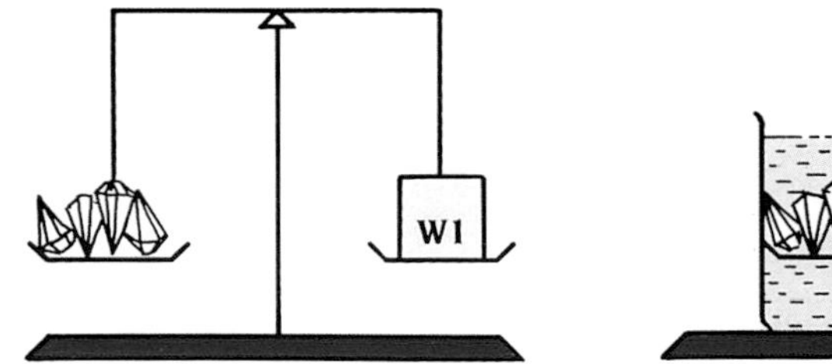

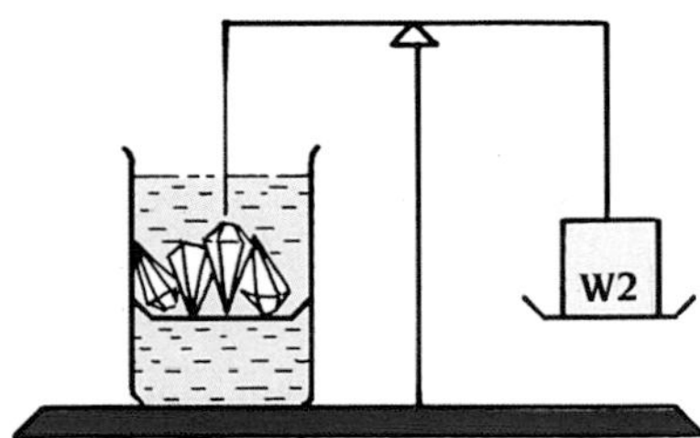

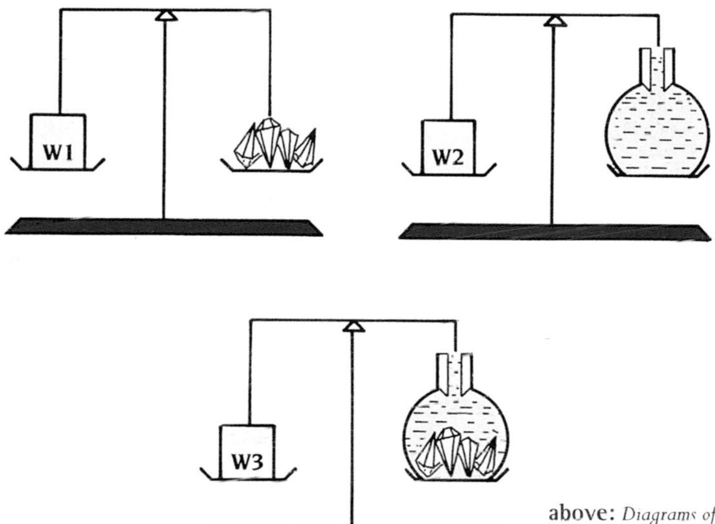

above: *Diagrams of the three stages of using a pycnometer.*

hydrostatic balance, the crystal is weighed in the air (W1) and then in distilled water (W2), and

$$\text{specific gravity} = \frac{W1}{W1-W2}$$

An alternative method uses a pycnometer or "density measurer." This is a bottle with a tight-fitting stopper with capillary holes in it. First the crystals are weighed on scales (W1), then the pycnometer is filled with distilled water, the stopper is put in place, and it is weighed (W2). Now the crystals are placed in the pycnometer, the stopper replaced, the excess water is removed, and the whole thing is weighed again (W3). In this case

$$\text{specific gravity} = \frac{W1}{W1+W2-W3}$$

The density of a material is a function of how closely together the atoms are packed. It is often used by mineral processing engineers as an easy and reliable way of separating useful minerals from worthless minerals.

Crystal Uses

Crystals were first used by primitive man as a cutting edge for weapons and tools, attached to a wooden shaft or handle. They were also used by cave dwellers to make paints. Examples of cave paintings still exist in southern France, northern Spain, and eastern Australia. The paints used were made from ground and powdered hematite (red), limonite (yellow), and pyrolusite (blue-black) mixed with either water or animal fats to a smooth paste. These preparations were also used as body paints in battle, at ceremonies, or for religious purposes.

Over 5,000 years ago in Ancient Egypt, the pharaohs forced thousands of slaves to work the great turquoise mines located within the Sinai Peninsula. These were perhaps the first commercially operated crystal mines. As well as turquoise, the Egyptians wanted lapis lazuli. The demand for this beautiful crystal could not be met by local production, so supplies were brought in from Afghanistan, about 2,500 miles away, and they even worked out how to make imitation crystal to meet deficiencies in supply.

Today, many commonplace items, such as plaster of Paris and rock salt, come from crystal sources. Crystals are used in the metals industry, not only as a source of most metal ores, but as the main ingredient of refractory bricks and the fluxes used in metallurgical refining. High-quality crystals continue to be used as gemstones.

A revolutionary step forward in crystal engineering came about as a direct result of the interest shown in ceramics as an alternative material to steel alloys. Ceramic materials and powdered metal are pressed into molds at such high temperatures and pressures that they fuse into a solid mass. This mass is machined to precise tolerances and may be used as, say, an engine block. The reduction in

above: *Crystals were often carried into battle. Here garnets adorn a silver belt buckle found in a sixth-century Viking grave in Norway.*

below: *This wooden box contains a selection of typical Egyptian jewelry and amulets, featuring such crystals as gold and lapis lazuli.*

right: *Long sought after for its beauty and endurance, gold is a rare metal which may be found in crystal form. This gold dredge is working a riverbed on South Island, New Zealand.*

weight and the degree to which the original powdered mass can be sculptured have been welcomed as a breakthrough in materials engineering. The next stage was the growing of a single crystal into the shape of an engine block or turbine blade using a heavily charged solution as the source medium. Work is continuing in this field with encouraging progress to date.

right: *Kennecott strip copper mine in Utah shows how dramatic this type of mining can be. The mined copper is widely used in the electrical engineering field.*

Synthetic Crystals

above: *A scene from the De Beers diamond sorting house.*

The Egyptians produced synthetic lapis lazuli by coloring alkali stones with blue copper cobalt. Not all imitations are designed to deceive and today synthetic crystals are an essential part of modern industrial practices.

In 1758 Joseph Strasser, a Viennese scientist, produced a type of glass that could be cut like a diamond. But it was in 1902 that the breakthrough came, when A. V. L. Verneuil, a French chemist, accidentally produced a synthetic ruby while he was trying to make a sapphire. The method he used is now called flame-fusion process. Powdered raw material is heated in a furnace to 3,632°F, the molten material drips through a hole, and is collected to form a candle-shaped mass, the "boule." This takes about four hours, then the boule is split lengthwise to relieve any internal stress. It is then ready to be cut into a crystal. By this method, sapphires, aquamarines, blue zircons, emeralds, and tourmalines can all be produced.

The first synthetic diamond was produced in 1955 from a graphite crystal. The crystal was subjected to such extremes of temperature and pressure that the carbon atoms rearranged themselves into a diamond. Development continued and the first crystal of gem quality was made in 1970. However the process is extremely expensive due to the fact that it requires the generation of pressure between 50,000 and 100,000 pounds per square inch at a temperature of 2,732 to 4,352°F. Despite the cost, synthetic diamonds which are not of gem quality play a huge role in industrial cutting, drilling, and grinding today.

Crystals are truly remarkable objects, usually formed an atom at a time, within a heated and high-pressure environment. It is astonishing that they exist at all. Irrespective of their appearance—perfect or imperfect, bright and beautiful, or dull and ugly—it is impossible to imagine a world without them.

below: *Brownish yellow synthetic diamonds manufactured at the De Beers Diamond Research Laboratory.*

Using The Crystal Identifier

The crystals in the identifier section are listed in descending order of hardness, using Mohs' scale. To find the reference for a specific crystal name, refer to the index at the back of the book.

Above each identifier entry is a group of symbols that indicate the hardness, specific gravity, crystal system (or classification), and main use of the crystal described. The symbols to indicate the use of the crystal are as follows:

Gemstone 

Ornamental 

Industrial 

Scientific 

Diamond

Mohs' Hardness 10

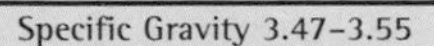

Specific Gravity 3.47–3.55

Crystal Structure Isometric

Its value as a gemstone is based on the "four Cs"; color, clarity, cut, and carat (weight). Only 20 percent of all diamonds are suitable for gemstones, with the rest being used in industry. Diamonds are formed at great depths by high temperature and pressure; then they are blasted by volcanic forces through vents toward the surface at such a speed that they are unable to cool into graphite, but form diamonds instead.

Some diamonds are colorless; others come in different tones of yellow—light, dull, or brown-tinted. In rare cases, some diamonds display a strong color (blue, brown, green, red, violet, or yellow) and are referred to as "fancy colored." A diamond's most distinctive features are its hardness and sparkle, which are unmatched by any other crystal.

Range: South Africa is the world's major supplier of gem-quality crystals, with the majority of industrial diamonds coming from deposits in Angola, Australia, and Zaire.

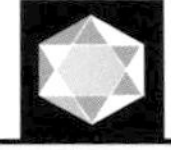

Ruby *corundum group*

Mohs' Hardness 9

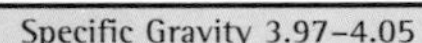

Specific Gravity 3.97–4.05

Crystal Structure Hexagonal

Rubies have long been prized as a precious gemstone. Lower quality crystals are powdered and used as a high-quality cutting and polishing medium. They tend to occur in dolomitic-type limestones which have become marble-like rocks. Most rubies are mined from alluvial deposits.

Rubies occur in a variety of shades of red with chrome being the coloring pigment; any brown hue which occurs is due to the presence of iron.

The distribution of color within the crystal is often uneven, occurring in irregular strips or spots. Ruby crystals become distinctly darker in color when they are exposed to natural light. They often possess a soft, silky sheen which is due to the inclusion of minute rutile crystals.

Range: The highest-quality rubies come from Burma, although Thailand supplies most of the world's markets. Small quantities of rubies also come from Afghanistan, Australia, Brazil, Cambodia, India, Malawi, Pakistan, and the United States.

Sapphire *corundum group*

Mohs' Hardness 9

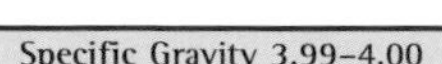

Specific Gravity 3.99–4.00

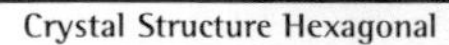

Crystal Structure Hexagonal

A sister crystal to ruby, the comparatively lower value of the sapphire reflects the relatively common occurrence of iron within the stone, from which it derives its distinctive blue color. Sapphires are formed as crystals in marble, basalt, and pegmatite, as a result of contact metamorphism between alumina-rich magma-type rocks and limestone.

Sapphire crystals occur in a variety of colors; black, purple, violet, dark blue, bright blue, light blue, green, yellow, and orange. Australian stones have a particularly deep blue color with distinctive blue-green pleochroism; Sri Lankan stones tend to have a patchy blue color, while Kashmir stones are a deep milky cornflower blue with no obvious inclusions.

Range: The most desirable crystals traditionally come from the Kashmir region of India but Australia is now the world's foremost source. Sapphires are also found in Brazil, Burma, Cambodia, Kenya, Malawi, Tanzania, Thailand, and Zimbabwe.

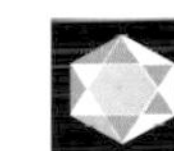

Chrysoberyl

Mohs' Hardness $8\frac{1}{2}$

Specific Gravity 3.7–3.72

Crystal Structure Orthorhombic

Green chrysoberyl, alexandrite (named after its discovery in the Urals in 1830 on the day Prince Alexander of Russia came of age), and honey-yellow cat's-eye, are all highly prized gemstones. The crystal is uncommon; it is usually associated with granite-pegmatites, mica schists, alluvial, or marine deposits.

Chrysoberyls can be green, yellow, gray, brown, or colorless. Alexandrite is very distinctive in that it is green in natural light and red in artificial light. Green grossular garnets display similar characteristics.

Cat's-eyes are greenish-yellow or yellow, often with a cold grayish tone; they display a moving light ray which along with their color lead them to resemble the eyes of feline creatures. Chrysoberyls tend to be sensitive to chemical attack by alkalis and can change color when heated.

Range: Chrysoberyls are found in Brazil, Burma, the CIS, Madagascar, Norway, Sri Lanka, Tanzania, the United States (Connecticut), and Zimbabwe.

Bixbite (or Red Beryl) *beryl group*

Mohs' Hardness $7^{1}/_{2}$–8	Specific Gravity 2.65–2.75	Crystal Structure Hexagonal

Due to its rarity and distinctive color, bixbite is a relatively highly valued semi-precious gemstone. While commanding a high market price, bixbite has not as yet been commercially imitated or produced synthetically. Unlike the other crystals of the beryl group which are found in or near pegmatic veins, bixbites are found in effusive magmatic rhyolite rocks.

Bixbite has a strong, ruby-red, violet, or strawberry-red hue. The crystals, which tend to be small, always contain numerous inclusions and more often than not internal flaws.
Range: Bixbite crystals are only found in the United States (New Mexico and Utah).

Emerald *beryl group*

Mohs' Hardness $7^{1}/_{2}$–8	Specific Gravity 2.67–2.78	Crystal Structure Hexagonal

The emerald has long been valued as a precious gem, those with deep green crystals commanding the highest prices. Emeralds are formed in the vicinity of rising magma and are characteristic of granites and pegmatites where they occur as crystals. Emerald crystals are bright green, light green, yellow-green, or dark green. The pigment is due to traces of chrome and sometimes vanadium. The color is very stable in light and heat, only altering at 1,292–1,472°F. Emeralds are only occasionally transparent; they mostly contain inclusions which are due to liquid or gas bubbles, heating cracks, or foreign crystals. These inclusions are considered to be a sign of authenticity and do not necessarily detract from the value of the crystal.
Range: The most important emerald source is Colombia, where the crystal—which occurs in calcite veins—was originally mined by the Incas. Emeralds also occur in deposits in Australia, Brazil, the CIS, India, Pakistan, South Africa, Tanzania, Zambia, and Zimbabwe.

Golden Beryl *beryl group*

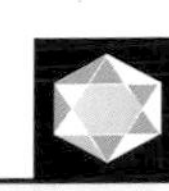

Mohs' Hardness 7½–8	Specific Gravity 2.65–2.75	Crystal Structure Hexagonal

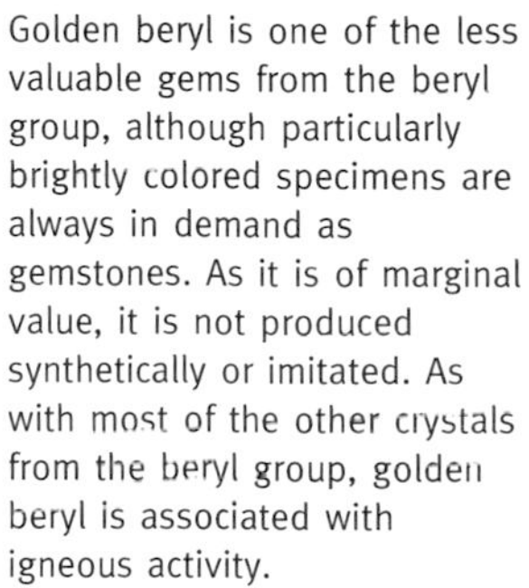

Golden beryl is one of the less valuable gems from the beryl group, although particularly brightly colored specimens are always in demand as gemstones. As it is of marginal value, it is not produced synthetically or imitated. As with most of the other crystals from the beryl group, golden beryl is associated with igneous activity.

The most highly prized golden beryl crystals range in color from canary yellow to golden yellow, while other stones are dull yellow or lemon yellow. The pigment is thought to be due to iron contained within the crystal structure. Inclusions are rare, when they do occur, they take the form of regular parallel bundles in a straw-like form, which—clearly visible with a lens—reduce the crystal's transparency and luster.

Range: Golden beryl comes primarily from the Minas Gerais state of Brazil, although other deposits occur in Madagascar, Sri Lanka, and the United States (Massachusetts and Virginia).

Morganite (or Pink Beryl) *beryl group*

Mohs' Hardness 7½–8	Specific Gravity 2.8–2.9	Crystal Structure Hexagonal

Morganite, which is named after the American banker and mineral collector John Morgan (1837–1913), is a strongly colored crystal which is much valued as a gemstone. Morganite crystals occur in or near pegmatite veins.

Morganite crystals are usually soft pink or violet in color, with no overtones. Lower color quality specimens can be heat-treated to 752–932°F where they improve into aquamarines. The crystals, which usually occur as long prisms, have a glassy luster. Although morganite crystals are usually free of inclusions, when they do occur, they are very irregularly shaped liquid or gaseous forms, which are only just visible.

Range: Important morganite deposits occur in the Minas Gerias region of Brazil, Madagascar, Mozambique, Namibia, the United States (California), and Zimbabwe.

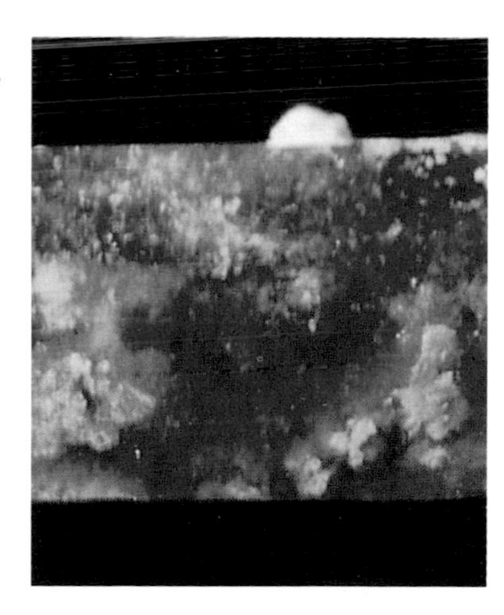

Phenacite (or Phenakite)

Mohs' Hardness $7^1/_2$–8	Specific Gravity 2.95–3.0	Crystal Structure Hexagonal

Quality transparent phenacite or phenakite, crystals are sometimes cut into gemstones. Phenacite has been produced synthetically on a limited basis. The crystals are formed in high-temperature pegmatite veins and in mica schists. They are frequently found with quartz, chrysoberyl, beryl, apatite, and topaz.

Phenacite crystals are colorless, or they can be white, yellow-tinted, or pink. Crystals are transparent with a vitreous luster. The crystal's cleavage is imperfect, but it is infusible and insoluble in acids.

Range: Phenacite crystals are found in Brazil, the CIS (the Urals), Mexico, and the United States (Colorado).

Andalusite

Mohs' Hardness 7–$7^1/_2$	Specific Gravity 3.12–3.2	Crystal Structure Orthorhombic

When available in large enough quantities, andalusite is used in industry for the manufacture of refractories, high-temperature electrical insulators, and acid-resistant ceramic products. Quality crystals are sometimes cut into gems, especially when they display a greenish or reddish color. Andalusite is a characteristic mineral of low-pressure metamorphic rocks (usually granitic or argillaceous). They are rich in aluminum and poor in calcium, potassium, and sodium.

The color of andalusite varies dramatically from light yellowish-brown to green-brown, light brownish-pink, bottle green, or grayish-green. Andalusite crystals have a modest luster and sometimes have dark inclusions running across the plane of the prism; these crystals are called chiastolite.

Range: Fine andalusite crystals are found in Spain in the vicinity of Andalusia (hence the name), as well as in Australia, Brazil, Burma, Canada (Quebec), Sri Lanka, and the United States (California, Maine, Massachusetts, and Pennsylvania).

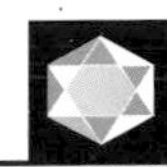

Danburite

Mohs' Hardness $7-7\frac{1}{2}$	Specific Gravity 2.97–3.02	Crystal Structure Orthorhombic

Danburite is of marginal value as a gemstone, despite its hardness. The crystals are usually found in fissures and in lining lithoclase, especially as an incrustation in albite.

Danburite crystals tend to be either colorless or pink; occasionally they are pale yellow. The crystals have poor cleavage and are transparent with a greasy luster. The boron in the crystal's make-up will color a flame green, and it can easily be fused into a colorless glass. The danburite crystal is insoluble in acid.

Range: It is named after Danbury (Connecticut, United States), where the crystal was first found. Other important deposits occur in the Swiss and Italian Alps, Japan, Madagascar, and Mexico.

Indicolite *tourmaline group*

Mohs' Hardness $7-7\frac{1}{2}$	Specific Gravity 3.02–3.26	Crystal Structure Hexagonal

Attractive clear blue, bright blue, or blue-green crystals of indicolite are often cut and set as gemstones. However, when the color is too deep or too blue, the stones fail to attract much interest from jewelers. Indicolite crystals commonly occur in greisen and pegmatites, where they have grown due to magmatic intrusions. Crystals also occur in sedimentary rocks as branch-like or authigenic grains.

Indicolite, named after the color indigo, is the blue variety of tourmaline, but crystals can also be a greenish-blue color. The intense color is a distinctive feature of indicolite, as is the loss of transparency when viewed in one direction. The crystals are insoluble in acid.

Range: Indicolite crystals are found in Brazil, the CIS (the Urals), Madagascar, Namibia, and the United States (California, Colorado, and Massachusetts).

Iolite (or Cordierite or Dichroite)

Mohs' Hardness 7–7½	Specific Gravity 2.53–2.66	Crystal Structure Orthorhombic

The blue variety of iolite, which is known as water sapphire, is cut into a moderately valuable gemstone. Iolite is formed in areas of contact metamorphism and less frequently in silica- and alumina-rich granitic or rhyolitic eruptive rocks.

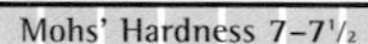

The name itself is derived from its sometimes violet color. The crystals are normally a variety of dark blue to light blue, with gray being the most common additional hue. There is also a black, iron-bearing iolite called sekaninaite.

The crystal has sometimes been referred to as the "Vikings' compass" in reference to its ability to indicate the direction of the sun on overcast days. It is sometimes confused with quartz; unlike quartz, however, it is fusible in thin sections and is insoluble in acid.

Range: Iolite crystals are mainly found in Brazil, Burma, Canada, Finland, India, Madagascar, Sri Lanka, Namibia, and the United States (Connecticut).

Pyrope (or Cape Ruby) *garnet group*

Mohs' Hardness 7–7½	Specific Gravity 3.65–3.87	Crystal Structure Isometric

Pyrope gets its name from the Greek word *pyropos*, meaning fiery. It is a moderately valuable semi-precious gemstone, with the darkest crystals being the most common. Pyrope is typically formed in peridotite and serpentinized periodotites, as well as in the diamond-bearing clay called kimberlite. Due to their resistance to weathering, pyropes are often found in alluvial or secondary deposits.

Pyrope is deep red in color, with the coloration due to traces of chrome in the crystal structure. Pyrope is singly refractive with birefrigence patches. Its luster is comparable to that of rubies and spinels. Pyrope fuses fairly easily and is almost insoluble in acid.

Range: Pyrope crystals are found in moderately sized deposits in Argentina, Australia, Brazil, the former Czechoslovakia, Mexico, South Africa, Tanzania, and the United States.

Rubellite *tourmaline group*

Mohs' Hardness 7–7½	Specific Gravity 3.02–3.26	Crystal Structure Hexagonal

Rubellite crystals which are ruby-colored and contain no inclusions are modestly priced and used as gemstones. Rubellites are usually formed in association with igneous and metamorphic activity. Crystals commonly occur in greisen and pegmatites but can also occur in sedimentary rocks as detrital and authigenic grains.

The color of rubellite crystals varies from pink to violet-pink, pink with a brown tint, to red or violet-red with a brownish tint. Violet-red rubellite crystals are sometimes referred to as siberite, after Siberia where they occur. The color is a distinctive feature of rubellite crystals. However, sometimes the color can be a little subdued. The crystals can also fail to become brighter in strong light, something they have in common with rubies.

Range: Rubellites are found in Brazil, Burma, the CIS (Siberia), Madagascar, Sri Lanka, and the United States (California).

Staurolite (or Fairy Stones)

Mohs' Hardness 7–7½	Specific Gravity 3.7–3.8	Crystal Structure Monoclinic

Staurolite (the word derives from the Greek *stauros* and *lithos*, meaning cross and stone) frequently occurs in cruciform-twinned crystals; the crystals either form a Greek cross (90 degrees between the arms) or a St Andrew's cross (60 degrees between the arms). Consequently, they are often worn as religious jewelry. Staurolite is a metamorphic mineral formed within a medium-temperature environment.

Staurolite crystals often have a rough earthy coating (due to surface chemical alteration) which hides their reddish-brown to black interior. They frequently occur embedded in kyanite or schist as flat elongated crystals. Crystals are generally semi-opaque, with a resinous luster. They are infusible but slightly soluble in sulfuric acid.

Range: Staurolite crystals are found in eastern Germany, Scotland, and Switzerland. The cruciform-twinned crystals regularly occur in France and the United States (Georgia, New Mexico, and Virginia).

Tourmaline

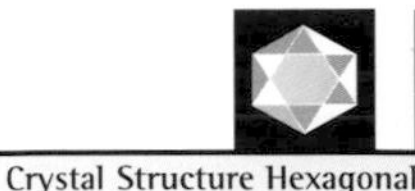

Mohs' Hardness 7–7½	Specific Gravity 3.02–3.26	Crystal Structure Hexagonal

Tourmaline was named from the Sinhalese word "toramalli" meaning high or hard rocks. Tourmaline will develop a positive and negative charge when rubbed or heated. High-quality, consistent tourmaline crystals are cut into gemstones; dramatically zoned stones often are sliced and polished as ornamental objects. Tourmalines either occur in intrusive dikes or silica-rich rock (especially granites and pegmatites), or as well-defined crystals due to their hardness.

Tourmaline occurs in a wide variety of colors; the most attractive are pink, fiery red, and deep green. Pale colored crystals can be heat-treated at 842–1,202°F to intensify coloration. Color zoning is common, and can vary from green at the base of a crystal to red at the apex. Crystals are brittle and are commonly elongated or columnar.

Range: The most widely occurring tourmaline is the black, iron-rich variety known as schorlite which is of little value. Sizeable and highly utilized deposits of other varieties of tourmaline occur in Brazil, Burma, the CIS (the Urals), Sri Lanka, Madagascar, Mozambique, and the United States (California, Connecticut, and Maine).

Verdelite (or Green Tourmaline) *tourmaline group*

Mohs' Hardness 7–7½	Specific Gravity 3.02–3.26	Crystal Structure Hexagonal

The name verdelite is derived from the Italian word meaning "green stone." Verdelite crystals range in color from light green to emerald green to deep green. Only clear crystals from the mid-range of colors are considered suitable for cutting into gems. The most distinctive features of verdelite crystals are their dark intense shades and the loss of transparency along the crystal axis. As with other crystals from the tourmaline group, verdelite crystals are associated with igneous, intrusive, and metamorphic environments, although they are commonly recovered from alluvial deposits.

Range: Verdelite crystals are found in Brazil, the CIS, Mozambique, Namibia, Sri Lanka, Tanzania, and the United States (Maine).

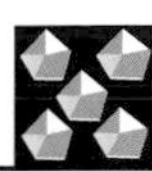

Amethyst *quartz group*

Mohs' Hardness 7	Specific Gravity 2.63–2.65	Crystal Structure Hexagonal

The name amethyst is derived from the Greek word *amethystos*, meaning "not drunk," as it was believed that by wearing the crystal one was protected from the effects of alcohol. It is the most highly prized form of quartz. However, its value fell at the turn of the century with the discovery of deposits in Brazil and Uruguay. High-quality crystals are still often used as a semi-precious gem. Amethyst crystals always grow from a base.

The pyramid-like crystals are not usually well developed and are therefore frequently found with the deepest color (violet, purple, or pink) at the tips. The color is due to traces of ferric iron. The distribution of varying color bands distinguishes amethysts from crystals of similar appearance. A crystal will turn white when heated to 572°F, and will yellow at 932°F.

Range: Some spectacular crystals come from the state of Minas Gerais in Brazil. Amethysts are also found in large quantities in Australia, Canada, the CIS, the former Czechoslovakia, India, Madagascar, South Africa, Sri Lanka, and the United States.

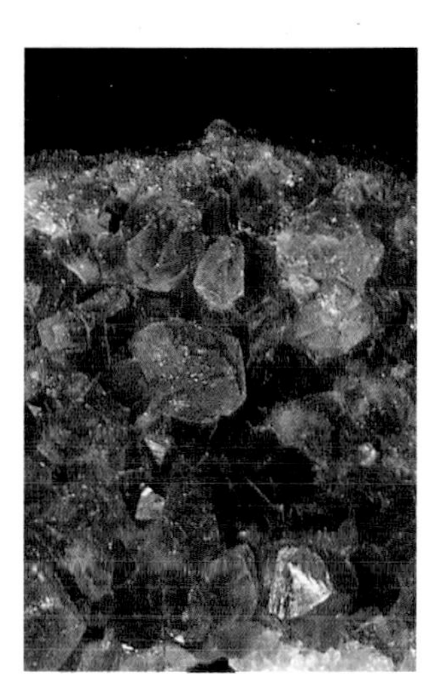

Citrine *quartz group*

Mohs' Hardness 7	Specific Gravity 2.65	Crystal Structure Hexagonal

Most commercial citrines are actually heat-treated amethysts or smoky quartz. Citrine is widely used as an imitation of the more expensive gemstone, topaz. The larger citrine crystals, which are prismatic with pyramid ends, are associated with intrusive magmatic phenomena.

The distinctive color is due to the presence of colloidal iron hydrates and varies from pure yellow to dull yellow, honey, or brownish-yellow. Citrine crystals will turn white if heated and dark brown if exposed to X-rays. As with amethysts, the color is often broken up into patches or bands. Cut crystals display good luster. The density of citrine is the lowest for crystals of this color.

Range: Citrine is rare but does occur in Brazil, the CIS, France, Madagascar, and the United States (Colorado).

Milky Quartz *quartz group*

Mohs' Hardness 7	Specific Gravity 2.65	Crystal Structure Hexagonal

Milky quartz is often cut into beads, ornaments, and *objets d'art*. The most common variety of milky quartz is found in pegmatites and hydrothermal veins; mystery still surrounds the process which leads to their formation and concentration.

The distinctive coloration of milky quartz is due to the inclusion of numerous bubbles of gas and liquid in the crystal. Milky quartz is the rougher, more compact formation of amethyst, layered and striped with milky bands.

Range: Milky quartz is one of the most common materials in the earth's surface (12 percent by volume), and occurs in many locations. Famous finds include a 14.5 ton crystal in the CIS (Siberia). It is especially common in central Europe, Brazil, Madagascar, Namibia, and the United States.

Rock Crystal (or Colorless Quartz) *quartz group*

Mohs' Hardness 7	Specific Gravity 2.65	Crystal Structure Trigonal

The name quartz comes from the Greek for ice, since it was once believed that the crystals were forever frozen by a process of extreme cold. Although quite common, rock crystal is often carved into *objets d'art* or fashioned into decorative jewelry. In the past it has been used for optical and piezoelectrical purposes; synthetic crystals are now generally used for this purpose. Rock crystal is crystallized directly from magma, in pegmatites, and in low-temperature hydrothermal regions.

Rock crystal is colorless, transparent, and—unlike glass—is birefringent. It is distinguishable from glass by its absence of air bubbles, and from lead-glass by its hardness (7 compared to 5).

Range: Quartz is one of the most commonly occurring minerals in the earth's crust (12 percent by volume). Brazil has produced some spectacular crystals which have weighed in excess of 4 tons.

Rose Quartz *quartz group*

Mohs' Hardness 7	Specific Gravity 2.65	Crystal Structure Hexagonal

Rose quartz is much valued as an ornamental material because of its attractive color and comparative rarity. However, it is not very popular because of its tendency to be brittle. Although rose quartz occurs in massive form in many pegmatites, well-formed crystals are rare.

The color varies from strong to pale pink, and appears to be caused by traces of manganese or titanium. The crystal is usually somewhat milky rather than perfectly transparent. Named after its color, the crystal is often cracked, and usually a little turbid. It is only in recent years that crystals with flat sides have been found. Crystals tend to lose their color when heated; they turn black when exposed to radiation.

Range: Rose quartz is not uncommon, but is usually found in compact masses. Quality crystals have been found in Brazil, Madagascar, and the United States (California and Maine).

Smoky Quartz (or Smoky Topaz) *quartz group*

Mohs' Hardness 7	Specific Gravity 2.65	Crystal Structure Hexagonal

Smoky quartz, with its intricate patterns, is often cut into gemstones or *objets d'art.* Crystals weighing up to 670 pounds have been found in hydrothermal veins in Brazil. Its distinctive smoky characteristic is probably due to rock crystal being subjected to natural radiation.

Smoky quartz is named for its smoky color. It can be brown, black, or smoky gray. Very dark crystals are called "morion." When heated to 572–752°F crystals turn yellow then white. Quality crystals will often contain rutile inclusions.

Range: Smoky quartz is found worldwide. Quality crystals have been found in Brazil, Madagascar, and in Alpine fissures.

Tiger-eye *quartz group*

Mohs' Hardness 7	Specific Gravity 2.64–2.71	Crystal Structure Hexagonal

Tiger-eye is often used for carving-boxes and other ornamental items as these will display its distinctive markings to their full advantage. Tiger-eye crystals are formed from fine fibrous quartz aggregates which have had the crocidolite (a type of hornblend) altered to a yellow color.

Tiger-eye crystals vary in color from gold-yellow to gold-brown stripes against an almost black background. The golden hue is due to the presence of brown iron. The fibers making up the stripes are concentrated into semi-parallel groupings.

Range: The most important tiger-eye deposit occurs in South Africa but it is also found in western Australia, Burma, India, and the United States (California).

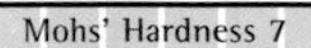

Quartz Cat's-eye *quartz group*

Mohs' Hardness 7	Specific Gravity 2.65	Crystal Structure Hexagonal

Despite its attractiveness, this material is not very valuable. It is usually cut into round polished pieces for necklaces or pendants. Quartz cat's-eye is formed from fluids associated with intrusive magmatic phenomena.

Quartz cat's-eye is semi-transparent but becomes greenish-gray or green when ground. The distinctive features of the crystals are their colors and clearly fibrous appearance. Often confused with chrysoberyl cat's-eye, the material is sensitive to acids.

Range: Quartz cat's-eye, which usually occurs in fibrous aggregates, is found in Burma, India, Sri Lanka, and western Germany.

Almandine *garnet group*

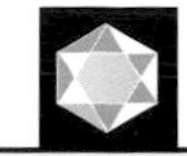

Mohs' Hardness $6^1/_2$–$7^1/_2$	Specific Gravity 3.95–4.2	Crystal Structure Isometric

Brightly colored almandine crystals which are free of inclusions and internal cracks are sometimes cut into gemstones. Almandine is often ground and used as a medium-hard abrasive in polishing paper and cloth. Almandine crystals, which are usually well-formed, are common in medium-grade metamorphic or contact metamorphic environments.

Almandine crystals have a red color but often contain a deep, violent-red tint. Although cut crystals have a brilliant luster, their transparency is often marred, even in clear stones, by an excessive depth of color. In an effort to lighten the stone the undersides of such gems are often hollowed out. Unlike rubies, the deep red color does not lighten in natural light. The crystal splinters and fuses easily, but is insoluble in acids.

Range: Almandine is found in large quantities in sand deposits in Sri Lanka, with lesser deposits occurring in Afghanistan, Brazil, the former Czechoslovakia, Greenland, India, Madagascar, Norway, Tanzania, and the United States (Alaska, California, Colorado, Connecticut, Idaho, Michigan, Pennsylvania, and South Dakota).

Grossular *garnet group*

Mohs' Hardness $6^1/_2$–$7^1/_2$	Specific Gravity 3.58–3.69	Crystal Structure Isometric

Green and yellow tinted grossular crystals are often cut and sold as gemstones. Usually associated with metamorphism, grossular crystals have usually been weathered from their host rock and are to be found in gem gravels.

Grossular crystals are frequently green, yellow, or copper-brown in color. Cinnamon-brown and orange grossular are not uncommon. The coloration is due to iron pigments, while the green variety is caused by chrome within the crystal structure. The name grossular is derived from the Latin word *grossularia* meaning gooseberries, which the crystals sometimes resemble. The crystals vary from semi-opaque to transparent, when they display good luster.

Range: Gem-quality grossular crystals are found in Canada, Kenya, South Africa, Sri Lanka (honey-yellow colored variety), Tanzania (green tinted variety), Madagascar, Mexico, and the United States.

Rhodolite *garnet group*

Mohs' Hardness $6^{1}/_{2}$–$7^{1}/_{2}$	Specific Gravity 3.74–3.94	Crystal Structure Isometric

Rhodolite is not the most common of the red garnets, but it is the most valuable. It is formed in plutonic and ultra-mafic rocks, but due to its resistance to weathering the crystals are usually found in alluvial secondary deposits or in arenaceous rocks.

The name rhodolite is derived from the Greek word *rhodon* meaning rose colored, and *lithos* meaning stone. The crystal itself can vary in color from pinkish-red to rose-red and pale violet. The cut crystal displays a strong luster and a good transparency. Rhodolite is distinguished from similar-colored crystals of the corundum group by its lack of pleochroism or fluorescence.

Range: Rhodolite crystal deposits are found in Brazil, Sri Lanka, Tanzania, Zambia, Zimbabwe, and the United States (North Carolina).

Spessartite *garnet group*

Mohs' Hardness $6^{1}/_{2}$–$7^{1}/_{2}$	Specific Gravity 4.12–4.20	Crystal Structure Isometric

Gem-quality spessartite crystals are extremely rare, but when they do occur, they are usually found unaccompanied by other such crystals and are well-formed. Spessartites are generally formed by, and associated with, low-grade metamorphic rocks.

Spessartite crystals are orange-pink, orange-red, red-brown or brownish-yellow in color. The crystals are unusual in that they display a high degree of hardness and density. They are semi-opaque or transparent; when transparent they have a high degree of luster.

Range: The name spessartite is derived from Spessart in western Germany, where the crystals were once found. The main deposits of spessartite occur nowadays in the gem gravels of Burma and Sri Lanka. Minor deposits are also found in Brazil, Madagascar, Mexico, Sweden, and the United States (California and Virginia).

Zircon

Mohs' Hardness $6\frac{1}{2}$–$7\frac{1}{2}$	Specific Gravity 3.9–4.71	Crystal Structure Tetragonal

Zircon is an important source of zirconium, hafnium, and thorium. Quality-grade crystals are cut into gemstones with green zircons being in much demand. Zircon crystals are usually associated with intrusive acidic igneous rocks or with pegmatites derived from them. They are often to be found in alluvial deposits as pebbles or grains.

Zircon crystals—which are usually four-sided and stubby in shape—range in color from colorless to yellow, red, brown, gray, and green. Crystals are sometimes perfectly transparent with good luster and strong birefringence, but they can also be opaque and dull. Zircon crystals are insoluble in acids; they are infusible but very brittle.

Range: Gem-quality crystals are found in Cambodia, Norway, Sri Lanka, Thailand, and Vietnam. Other important zircon deposits also occur in Australia, Brazil, the CIS (the Urals), and the United States (Florida).

Agate *quartz group*

Mohs' Hardness $6\frac{1}{2}$–7	Specific Gravity 2.60–2.65	Crystal Structure Hexagonal

Agate is the name given to microcrystalline quartz which is banded. These bands can be multicolored, or as is more usual, different shades of the same color.

Agates are nodules or geodes, ranging in size from a fraction of an inch to several feet in radius. They are formed from gas-created voids which have become filled with silica in a volcanic environment. The coloration is due to minute quantities of varying elements, while the banding is due to the gradual cooling of the material. Agate nodules usually have a white outer layer that is due to weathering.

Agates, and their banding, come in every conceivable color, shade, and hue. It is often a question of personal taste which agates are the more valuable; generally, those with regular widths of banding, crisp delineation between bands, and sharp vivid colors tend to appeal to both collectors and jewelers. Agates are usually cut and polished; if struck, they have a tendency to chip or splinter.

Range: The most important agate deposits occur in Brazil, India, and Uruguay; lesser agate deposits occur in Canada, the CIS, Germany, and the United States.

Peridot (or Olivine or Chrysolite)

Mohs' Hardness 6½–7	Specific Gravity 3.27–4.20	Crystal Structure Orthorhombic

Good-quality, clear colored crystals are often cut and set with other gemstones. Peridot is very widely distributed in iron- and magnesium-rich igneous rocks which form a continuous series. As the amount of iron contained within the crystal structure increases its specific gravity, its solubility increases and its fusing point lowers.

Peridot crystals are typically olive green, bottle green, yellowish-green, or brown in color. Crystals display a greasy vitreous luster when split, are usually transparent with few inclusions, and are not resistant to sulfuric acid. Dark crystals can be lightened in color by being burned.

Range: Peridots are found as large crystals on the island of Zebirget in the Red Sea, in the basalt formations in the United States, and in certain lavas on the Hawaiian Islands, Australia, Brazil, and South Africa (where they are found alongside diamonds).

Tanzanite (or Blue Zoisite)

Mohs' Hardness 6½–7	Specific Gravity 3.11–3.40	Crystal Structure Orthorhombic

Good-quality tanzanite crystals are much sought after by jewelers. The crystals are formed as a result of the metamorphism of plagioclases. Larger crystals are formed in high-pressure and high-temperature environments or in hydrothermal veins often associated with sulfides.

In good-quality tanzanite crystals, the color varies from ultramarine to sapphire blue, with amethyst-color and violet also occurring. The coloration is due to the presence of chromium and strontium within the crystal structure. When heated to 752–932°F, yellow and brown tints disappear and blue hues deepen. Crystals, which are usually transparent with vitreous luster, are insoluble in acid and fuse relatively easily into a white blister-like glass.

Range: It is found in limited quantities in Tanzania (hence its name); these sources are now almost exhausted.

Vesuvianite (or Idocrase)

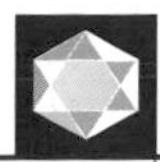

Mohs' Hardness $6\frac{1}{2}$	Specific Gravity 3.27–3.45	Crystal Structure Tetragonal

Good-quality green vesuvianite crystals (called californite) are popular with jewelers. They occur sometimes in calcareous blocks from mafic ejections, but are normally a product of contact metamorphism.

Vesuvianite crystals occur in a variety of colors from brown to green (californite), olive green, very occasionally yellow (xanthite), blue (cyprine), and red or white (wiluite). Crystals are usually opaque, but translucent and transparent varieties of vesuvianite usually have a vitreous to resinous luster. Crystals fuse easily and are virtually insoluble in acid.

Range: Vesuvianite was first found on Mount Vesuvius in Italy, hence the name. Californite comes from Pakistan and the United States (California), cyprine is found in Norway, wiluite comes from the CIS, while xanthite deposits occur in the United States (New York State).

Cassiterite

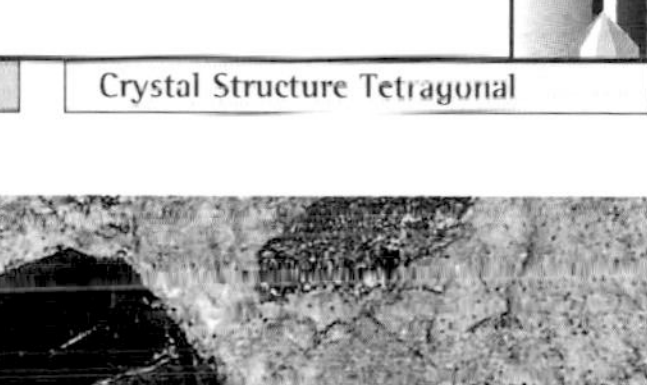

Mohs' Hardness 6–7	Specific Gravity 6.8–7.1	Crystal Structure Tetragonal

Named after the Greek word for tin, cassiterite is an important tin ore. The crystals, which are short and prismatic, typically occur in pegmatites and greisens. The large industrial-type deposits are usually sedimentary in nature. They were formed in fluvial or marine environments, and now occur as placers. Deposits are often worked by panning, as gold often is; in this way, the heavy crystals are separated from the light material in which they are found.

Cassiterite crystals are normally brown to black in color, but can sometimes be colorless or pink brown. Crystals display a brilliant luster and are infusible and insoluble in acids.

Range: The largest cassiterite deposits occur in Bolivia, Brazil, China, the CIS, Malaysia, and Sumatra.

Epidote (or Pistacite)

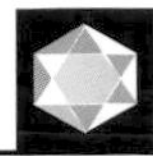

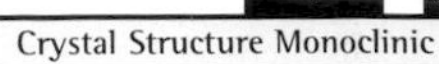

Mohs' Hardness 6–7 | Specific Gravity 3.3–3.5 | Crystal Structure Monoclinic

Epidote is sometimes polished and used for inlay work; other crystals are used occasionally as gemstones. Epidotes are formed in regional or contact metamorphic rocks of mafic composition.

Epidote crystals vary in color from full green to yellow or brown-black. The variety which is cherry-red to purplish-brown in color contains traces of manganese, and is known as piemontite. The gray, pale green, green-brown, or pink variety contains only minute quantities of iron and is called clinozoisite. This frequently occurs as a secondary hydrothermal alteration. Epidote crystals are transparent with a vitreous luster and fuse fairly easily.

Range: Crystal deposits are found in Austria, Norway, and the United States.

Hiddenite *spodumene group*

Mohs' Hardness 6–7 | Specific Gravity 3.16–3.20 | Crystal Structure Monoclinic

The scarcity of reasonably sized, attractive crystals makes the intensely colored hiddenite crystal a valuable, semi-precious gem. Crystals, which are usually long and unevenly terminated, are formed in lithium-bearing pegmatites associated with quartz, feldspar, beryl, and tourmaline.

Hiddenite is the green variety of spodumene, named after W. E. Hidden, the owner of the mine in the United States (North Carolina) where the crystal was first discovered in 1879. The color of hiddenite crystals ranges from pale green to yellow-green, and from emerald green to rich green. Hiddenite can resemble a number of other crystals (beryl, chrysoberyl, diopside, emerald, or verdelite) depending on its color. However, by testing and comparing the crystals' physical properties it should be possible to achieve a positive identification.

Range: The finest examples of gem-quality crystals are found in the United States (North Carolina) while less attractive, paler crystals occur in Brazil, Burma, Madagascar, and the United States (California and North Carolina).

Kunzite *spodumene group*

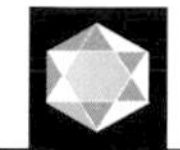

Mohs' Hardness 6–7	Specific Gravity 3.16–3.20	Crystal Structure Monoclinic

Kunzite crystals with few inclusions and good transparency are often cut into gemstones. Kunzite, which is an important source of lithium and its salts, is formed in lithium-bearing pegmatites associated with quartz, feldspar, beryl, and tourmaline.

Kunzite is the pink-violet, light violet, green-violet, or brown variety of spodumene. The crystal is named in honor of George F. Kunz (1850–1932), the mineral collector who first described it at the turn of this century. The kunzite crystals, which are usually long and unevenly terminated, generally have few inclusions. They are transparent, and display a marked pleochroism which is seen as a difference in color depth in different directions.

Range: Crystals are found in Brazil, Madagascar, and the United States (California, Connecticut, and Maine).

Amazonite (or Amazon Stone) *feldspar group*

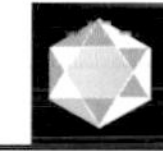
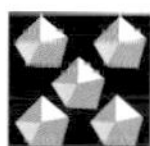

Mohs' Hardness 6–6½	Specific Gravity 2.56–2.58	Crystal Structure Triclinic

Amazonite, which is frequently confused with jade or turquoise, is often ground into beads for necklaces or fashioned into ornamental objects. Amazonite crystals, which are normally squat and slightly prismatic, are found in metamorphic, intrusive, and pegmatic rocks.

Deriving its name from the Amazon River, and the belief that the crystals somehow flowed from it, the amazonite crystal is usually light green but sometimes blue-green or bluish. Amazonite usually has a mottled appearance and sometimes has a fine criss-cross network of light striations. Crystals are generally semi-opaque, with poor luster and easy cleavage.

Range: Important amazonite deposits occur in Australia, Brazil, the CIS, India, Madagascar, Namibia, the United States, and Zimbabwe.

Benitoite

Mohs' Hardness 6–6½	Specific Gravity 3.65–3.68	Crystal Structure Hexagonal

Benitoite crystals, which are sometimes confused with light-colored sapphires, are often cut into gemstones. The crystals are found in veins in the brecciated (or fragmented) body of a blue schist associated with serpentinite.

Benitoite crystals, which are usually stubby and zoned, can range in color from light to dark blue. Crystals look blue when viewed through the acute faces of the rhombohedron and colorless when viewed through the obtuse faces.

Range: Benitoite is named after the only deposit found to date, which occurs in San Benito county, California. However, modern exploration techniques make it likely that it will not be long before other deposits are discovered.

Labradorite (or Spectrolite) *feldspar group*

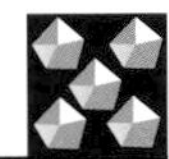

Mohs' Hardness 6–6½	Specific Gravity 2.62–2.76	Crystal Structure Triclinic

Labradorite is usually fashioned into decorative boxes or *objets d'art*, where its colors can be displayed to their full advantage. Smaller specimens are sometimes made into beads, brooches, or ring stones. Labradorite is typically associated with eruptive and metamorphic rocks.

Labradorite may at first appear to be a dark smoke-gray color, but when the light strikes it in a certain way, it displays rainbow-like colors (violet, indigo, blue, green, yellow, orange, and red) in an effect similar to that of gasoline lying on water. The most distinctive feature of labradorite is its iridescence against a dark background. This effect is probably caused by the interference of light on twinned lamellae (or layers).

Range: Labradorite usually occurs as a compact aggregate and rarely as an individually formed crystal. The world's most spectacular specimens come from Finland, with lesser rocks coming from Canada, the CIS, Madagascar, Mexico, and the United States.

Moonstone (or Adularia Moonstone) *feldspar group*

Mohs' Hardness $6-6\frac{1}{2}$	Specific Gravity 2.56–2.62	Crystal Structure Monoclinic

Moonstones, which display a blue reflection, are highly prized by jewelers. They could be confused however, with heat-treated amethysts or with milky synthetic spinels, if it were not for the fact that these two crystals do not display the correct mobile reflective characteristics. Moonstone is a variety of feldspar formed in association with orthoclase and albite (with a predominance of orthoclase).

In general, moonstones are almost colorless, having only a pale gray or yellow tint, with a whitish to silvery-white or blue shimmer. Incipient cleavage cracks are sometimes visible within the crystals; they have a slight turbidity and a distinctive mobile reflection.

Range: Important deposits of moonstone occur in Australia, Burma, India, Sri Lanka, Tanzania, and the United States.

Prehnite

Mohs' Hardness $6-6\frac{1}{2}$	Specific Gravity 2.87–2.95	Crystal Structure Orthorhombic

Prehnite is of interest to geologists as an indicator of formation sequences. Impressive crystals of prehnite can be found in museums and in mineral collections. Prehnite crystals have usually crystallized out from hydrothermal fluids in cavities contained within basaltic volcanic rocks.

Prehnite crystals are rare but when they occur they are usually white-green, light green, green-yellow, or yellow-brown. They can also occur as similar colored stalactitic aggregates. The crystals display perfect basal cleavage and are transparent with vitreous luster. They fuse easily, and dissolve slowly in hydrochloric acid with no gelatinous-silica residue.

Range: Found in Australia (New South Wales), China, France, South Africa, and the United States (Lake Superior, New Jersey, and Virginia).

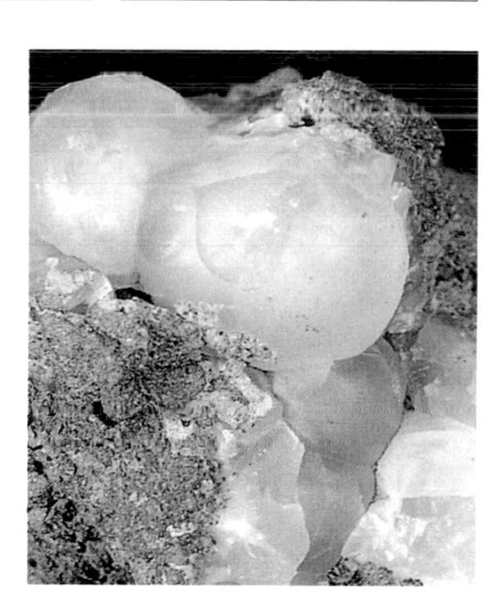

Opal (or Precious Opal) *quartz group*

Mohs' Hardness $5^1/_2$–$6^1/_2$	Specific Gravity 1.98–2.20	Crystal Structure Non-crystalline

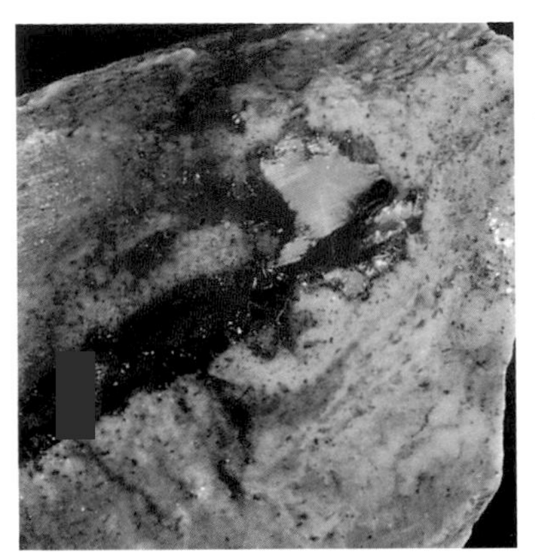

Opal is used as a gemstone because it contrasts well with other items of jewelry. It is never found in visible crystals, although recent work using electron microscopes has shown that the opal consists of minute balls of the mineral cristobalite suspended in a jelly-like matrix of silica. Opals occur as fine-grained masses occupying voids and veins, usually in a grape-like shape.

Opals can be divided into two groups: white or light-colored opals, which are known as white or milky opals; and the rarer black opals which can range in color from dark green to dark gray, and from dark blue to gray-black or totally black. Opals are extremely porous, and since they can contain up to 30 percent of water by volume, they age quickly by giving up this water. They fracture and chip easily, and are super-sensitive to acids, alkalis, and heat.

Range: Opals have been mined in the former Czechoslovakia since Roman times, and until the nineteenth century this was the only source of the stone. Today most of the high quality opals come from the Lightning Ridge area of New South Wales (Australia). Opals also occur in Guatemala, Honduras, Mexico, and the United States (California, Idaho, Nevada, Oregon, and Wyoming).

Rhodonite

Mohs' Hardness $5^1/_2$–$6^1/_2$	Specific Gravity 3.4–3.7	Crystal Structure Triclinic

Rhodonite is often used to make ornamental figures and large *objets d'art*; these are particularly valuable when veined by black manganese oxide. Rhodonite is formed through the metamorphic action of impure limestones which are themselves the result of element substitution involving magmatic fluids.

Rhodonite crystals tend to be pink, reddish-pink, or brown with veins or patches of black caused by the presence of manganese oxides. Rhodonite crystals display perfect prismatic cleavage (almost at right angles), and are transparent to translucent with a vitreous luster. Rhodonite is insoluble in acid, unlike the similar but softer rhodochrosite.

Range: Rhodonite is fairly widespread; quality crystals are found in Australia, Brazil, the CIS (the Urals), Finland, India, Japan, Madagascar, Mexico, New Zealand, South Africa, and the United States.

Sodalite

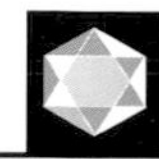
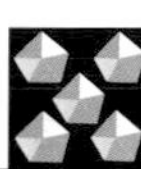

Mohs' Hardness $5^1/_2$–6	Specific Gravity 2.13–2.33	Crystal Structure Isometric

Sodalite is used in bead necklaces and *objets d'art*. It can easily be confused with lapis lazulite, especially as pyrite occurs in both. Sodalite occurs in under-saturated plutonic rocks, associated with metamorphosed limestones and volcanic blocks.

The color of sodalite varies greatly from bright blue to violet, and from white to gray with green tints. The white is usually due to calcite. Sodalite crystals are fragile with poor cleavage, and translucent with vitreous luster. They are soluble in hydrochloric and nitric acid and leave a silica gel when dissolved. Because of the presence of sodium, the crystal fuses easily giving a yellow flame.

Range: The most important sodalite deposits are in Brazil (Bahia) and Canada (Ontario). Less substantial deposits occur in Bolivia, Burma, the CIS, Greenland, India, Portugal, Romania, and Zimbabwe.

Lapis Lazuli (or Lapis or Lazulite)

Mohs' Hardness 5–6	Specific Gravity 2.38–2.9	Crystal Structure Isometric

Lapis lazuli has been used as an ornamental stone for thousands of years; it is often fashioned into large *objets d'art* or else cut into slabs for facing purposes.

During the Middle Ages, lapis lazuli was often ground to a powder and used as a pigment; today this pigment (produced from synthetic stone) is used in ultramarine paint. Lapis lazuli is an uncommon mineral, which usually occurs as compact masses associated with metamorphosed limestone (marble).

The name lapis lazuli is a hybrid term deriving from both Latin and Arabic meaning "blue-stone" (the stone has a distinctive azure hue). Lapis lazuli is a complex mineral and consists of a combination of several minerals, including calcite, augite, mica, pyrite, and pyroxene. Although it polishes well, it is fragile and will break imperfectly. It is porous and may be treated with paraffin to enhance the distinctive blue color.

Range: The best quality and most highly prized lapis lazuli comes from Afghanistan. Chile and Russia also produce large quantities of the stone, while small deposits also occur in Angola, Burma, Canada (Labrador), Pakistan, and the United States (California and Colorado).

Brazilianite

Mohs' Hardness 5½	Specific Gravity 2.98–2.99	Crystal Structure Monoclinic

Brazilianite is a rare and beautiful crystal which is much prized by jewelers and collectors. The crystals occur in cavities in pegmatites associated with blue apatite, clay, and lazulite.

Brazilianite crystals, which are usually elongated and stubby prisms, are yellow or green-yellow in color. They are often quite large. The crystals are fragile with perfect cleavage, transparent with a vitreous luster. They will dissolve with difficulty in strong acids, and will only fuse in small fragments. They quickly lose their color when heated.

Range: As the name suggests, Brazil is the major, and was until relatively recently the only, source of this crystal. It has now been found in the United States (New Hampshire) as well.

Chromite

Mohs' Hardness 5½	Specific Gravity 4.5–4.8	Crystal Structure Isometric

Chromite is the most important ore for chrome. Due to its hardness and resistance to chemical attack, it is used for plating and alloying with other metals. This produces hard chrome steel, stainless steel, and nickel-chrome. Chromium is used in the dying and tanning of leather, for producing yellow pigment, and plays a significant part in the manufacture of refractory bricks. It occurs as a primary mineral of ultra-basic rocks and sometimes in deposits which have been transported and later deposited in a different location.

Chromite rarely occurs as individually formed crystals, but more commonly as a brown-black or black submetallic granular mass. It is weakly magnetic, infusible, and insoluble in acids.

Range: Chromite occurs in quantities worthy of economic extraction in the CIS (the Urals), Cuba, India, New Caledonia, South Africa, and Turkey.

Enstatite *pyroxene group*

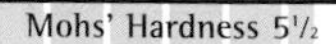

Mohs' Hardness 5½ | Specific Gravity 3.26–3.28 | Crystal Structure Orthorhombic

Enstatite is of interest to geologists, scientists, and crystal collectors alike, as its resistance to melting means it is often a good indicator of the direction of original lava flow. It is formed in mafic and ultramafic, plutonic, and extrusive volcanic environments, as well as in high-grade metamorphic rocks.

Enstatite crystals are sometimes stubby and prismatic; more often, they are fibrous or plate-like masses. They come in a variety of colors, ranging across yellowish, green, olive green, gray-green, or gray. The green-brown variety has a high iron content and is called bronzite. The name enstatite comes from the Greek word meaning very resistant to melting, and as one would expect it is insoluble in acids and almost infusible. Crystals display good cleavage; they are transparent with a vitreous luster changing to pearly on cleavage surfaces.

Range: Enstatite is found in the CIS, Germany, India, Japan, Northern Ireland, Norway, Scotland, South Africa, and the United States.

Nickeline (or Niccolite)

Mohs' Hardness 5–5½ | Specific Gravity 7.5–7.8 | Crystal Structure Hexagonal

Nickeline was the first mineral from which nickel was derived. It is still employed for this purpose when sufficiently large masses occur. Nickel is of major importance as an alloying metal. It is used in German silver (nickel, copper, and zinc), nichrome (nickel, chrome, and iron), nickel steel (nickel and iron), money metal (25 percent nickel and 75 percent copper), and stainless steel. Nickel alloys possess a number of useful characteristics: they are ductile, strong, resistant to chemical attack, and heat stable.

Nickeline occurs in high-temperature hydrothermal veins.

Nickeline crystals are extremely rare, but, when they do occur, they are squat, plate-like, or pyramidal. More frequently, they occur as a compact metallic bronze mass with pink, blue, gray, and sometimes turquoise colorations. It fuses easily giving off a strong, garlic-like smell. Nickeline is soluble in nitric acid producing a green solution.

Range: Crystals have been found in Germany, while large masses of nickeline occur in Argentina, Canada, Germany, and Japan.

Diopside *pyroxene group*

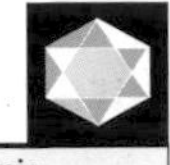

Mohs' Hardness 5–6	Specific Gravity 3.27–3.31	Crystal Structure Monoclinic

Despite the fact that diopsides are often confused with emeralds, hiddenites, or peridots, they are popular with jewelers; transparent specimens often are cut into gems. Diopside crystals are formed from contact metamorphism, especially in dolomitic marbles which are associated with other calcium silicates.

Diopside crystals are usually white, yellow, green, blue, brown or, very occasionally, colorless. The purple manganese-bearing variety is known as violane; the dark green chromium-bearing variety is known as chromian diopside; and the vanadium-bearing variety is known as lavrovite. Crystals are fragile and have perfect prismatic cleavage. They are transparent to translucent with a vitreous luster. They have a tendency to fuse easily but are insoluble in acids.

Range: Diopside crystals are found in Austria, Finland, Greenland, Madagascar, South Africa, and Sweden.

Lazulite

Mohs' Hardness 5–6	Specific Gravity 1.61–1.65	Crystal Structure Monoclinic

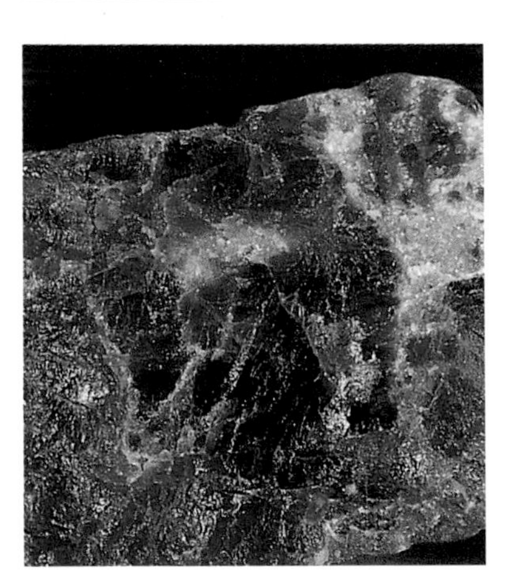

Lazulite is a gemstone of little importance. It is found in silica-rich rocks, pegmatites, and in quartz veins associated with andalusite and rutile or metamorphic rocks.

Lazulite crystals, which are rare, can be dark blue, bright blue, or blue-white. They possess a distinctive prismatic cleavage and are transparent with a vitreous luster. They will dissolve with difficulty in hot concentrated acid, are infusible, but will discolor and crumble when heated.

Range: Lazulite is found in Austria, Brazil, India, Madagascar, Sweden, Switzerland, and the United States (California, Georgia, Maine, and North Carolina).

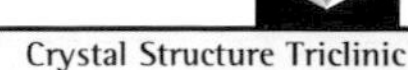

Turquoise

Mohs' Hardness 5–6	Specific Gravity 2.6–2.9	Crystal Structure Triclinic

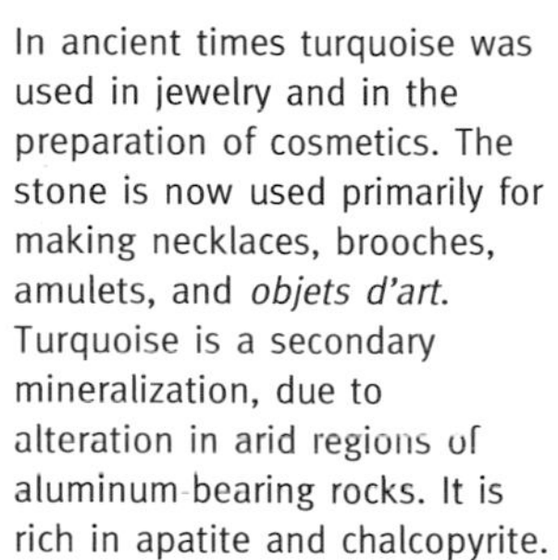

In ancient times turquoise was used in jewelry and in the preparation of cosmetics. The stone is now used primarily for making necklaces, brooches, amulets, and *objets d'art*. Turquoise is a secondary mineralization, due to alteration in arid regions of aluminum-bearing rocks. It is rich in apatite and chalcopyrite.

Turquoise comes in a range of colors from blue-white to brilliant sky-blue, and light greenish blue to light blue; it is generally opaque. Surprisingly, turquoise was not thought to be crystalline until crystals were found in Virginia, in the United States, in 1911. Sky-blue crystals turn dull green when heated at 482°F. Turquoises, which usually occur as grape- or kidney-shaped aggregates, are soluble in hydrochloric acid.

Range: Turquoise means "Turkish-stone," and is so called because the stones used to be brought to Europe along the overland trading route through Turkey. The best quality turquoise comes from Iran, with prolific deposits

being worked in the United States (Arizona, California, Nevada, and New Mexico).

Uraninite (or Pitchblende)

Mohs' Hardness 5–6	Specific Gravity 7.5–10	Crystal Structure Isometric

Uraninite is the main source for uranium and radium. The Curies (Marie (1867–1934), and her husband, Pierre (1859–1906)) first identified polonium, radium, and helium, from a sample of pitchblende. These elements were known even then to occur in the sun, but had still not been isolated and identified back on earth. Uranium is used in power generation, in nuclear fusion, and as a source of radium whose radioactive qualities are used in many branches of security. It has many uses in medicine, and in monitoring (fluid density, speed, quantity, etc). Uraninites often occur in pegmatites, medium- or high-temperature hydrothermal veins, or concentrated in gold-bearing conglomerates due to their high specific gravity.

Uraninite occasionally occurs as a dull black cubic crystal, but more frequently is found as a granular mass or aggregate which commonly displays a bright yellow or orange color, or combinations of the two. It is a very heavy mineral, fragile but with no predictable cleavage patterns; is highly radioactive, infusible, and will dissolve in most acids except hydrochloric acid.

Range: Uraninite is mined in Australia, Canada, France, Namibia, South Africa, and the United States (Arizona, Colorado, and Utah).

Apatite

Mohs' Hardness 5 | Specific Gravity 3.16–3.23 | Crystal Structure Hexagonal

Apatite is used in the manufacture of phosphate fertilizers and in the chemical industry to make salts of phosphoric acid and phosphorous. It is associated with all types of eruptive rocks, in some hydrothermal veins, and in iron-rich igneous rocks. It is also common in marine sedimentary rocks where it has been formed by chemical deposition.

Apatite crystals vary in color from colorless to yellow, from green to brown, and occasionally from red to violet or blue. The crystals are very fragile with poor cleavage parallel to the base. They tend to be transparent to opaque with a vitreous luster. Some crystals lose their color when heated while others will fluoresce a bright yellow in ultraviolet light.

Range: Apatite is a common material, found in Austria, Canada (Ontario), Chile, Mexico, Morocco, Nauro, South Africa, Togo, Tunisia, and the United States (Florida and Maine).

Dioptase

Mohs' Hardness 5 | Specific Gravity 3.28–3.35 | Crystal Structure Hexagonal

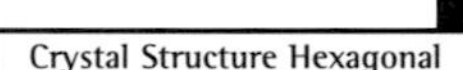

Dioptase is used for making jewelry and is popular with crystal collectors. It occurs as short prismatic crystals within cavities in the oxidation zones of copper deposits.

Dioptase crystals are a bright emerald-green color. They are fragile with perfect cleavage, transparent to translucent with a vitreous luster. Although infusible, they will expand and turn black when heated. The crystals are soluble in ammonia; they dissolve in hydrochloric acid and nitric acid, leaving a silica residue.

Range: Dioptase deposits are found in Chile, the CIS, Namibia, the United States (Arizona), and Zaire.

Hemimorphite

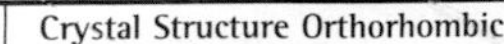

Mohs' Hardness $4\frac{1}{2}$–5	Specific Gravity 3.4–3.5	Crystal Structure Orthorhombic

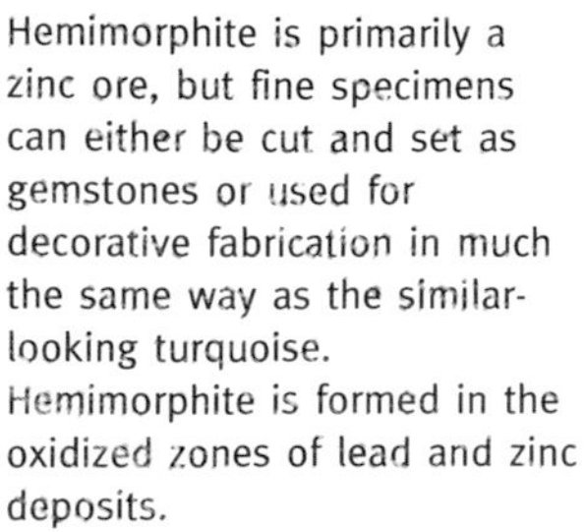

Hemimorphite is primarily a zinc ore, but fine specimens can either be cut and set as gemstones or used for decorative fabrication in much the same way as the similar-looking turquoise. Hemimorphite is formed in the oxidized zones of lead and zinc deposits.

Hemimorphite is rarely found as large crystals, but more commonly as platex crystals whose ends are different or hemimorphic (hence the name). Crystals, when they do occur, tend to be white, transparent, or translucent, but are more often tinted a yellow, green, blue, or brownish hue by the presence of copper or iron. In North America, hemimorphite is often referred to as calamine while in Europe smithsonite is often called calamine. It is soluble in strong acids but fuses with difficulty.

Range: Hemimorphite deposits are found in Algeria, Greece, Italy, Mexico, Namibia, and the United States (Colorado, Montana, and New Jersey).

Scheelite

Mohs' Hardness $4\frac{1}{2}$–5	Specific Gravity 5.9–6.1	Crystal Structure Tetragonal

Scheelite is an important source for tungsten, which is used for alloying in order to produce a very strong steel. Scheelite crystals are formed in high-temperature pegmatitic and hydrothermal veins.

Scheelite crystals range in color from no color at all to yellow (due to molybdenum traces), and from orange to brown. The crystals, which tend to be striated, are fragile with good cleavage and tend to be translucent or transparent with vitreous luster. Scheelite is soluble in acids and fuses with difficulty.

Range: Where industrial exploitation is concerned, the most important deposits occur in Australia, Bolivia, Burma, China, Japan, Malaysia, and the United States.

Platinum

Mohs' Hardness 4–4½ | Specific Gravity 14.0–19.0 | Crystal Structure Isometric

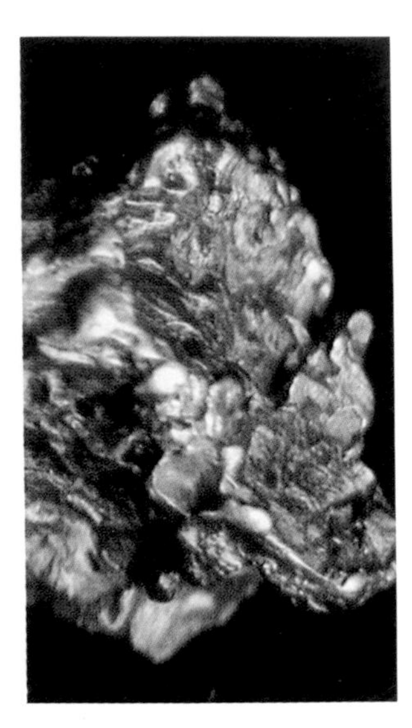

When the electronics industry was expanding in recent decades, platinum's ability to conduct electricity without oxidizing led to a renewed interest in this neglected metal. Interest grew when it was realized that there were many uses to which platinum (and its associated platinum group metals—or PGMs) lent itself as a chemical catalyst. Today, the world's need for PGMs as catalytic exhausts has contributed greatly to the unprecedented demand for platinum.

Platinum is formed by the action of hydrothermal fluid intrusion in basic and ultra-basic igneous rocks; it is also found in depositional areas derived from these rocks. The crystal usually occurs in rounded grains or scales, and very occasionally as nuggets. It is malleable with a metallic silver-gray luster. Platinum is very heavy and may be weakly magnetic. It will not dissolve in acids (except aqua regis) and only fuses at 3,227°F.

Range: The world's major source of platinum is the Witwatersrand region of the Transvaal (South Africa). Smaller quantities of platinum are found in Australia, Canada (Ontario), the CIS (the Urals), and the United States (Alaska).

Kyanite (or Disthene)

Mohs' Hardness 4–5 along axis, 6–7 across axis | Specific Gravity 3.65–3.69 | Crystal Structure Triclinic

Kyanite is a major raw material that is used for industrial purposes required for the manufacture of high-temperature porcelain products, perfect electrical insulators, and acid-resistant products. Formed in perlitic rocks rich in aluminum and metamorphosed under high pressure, kyanite crystals are usually long, flat, and prismatic.

Kyanite crystals vary considerably in color, from colorless to gray, and from green to blue-green or blue. Crystals are fragile, display perfect cleavage, and are transparent or translucent, with a pearly luster on cleavage planes. Crystals are infusible and are insoluble in acid.

Range: Kyanite is found in Australia, Austria, France, India, Kenya, Switzerland, and the United States (Connecticut, Massachusetts, and North Carolina).

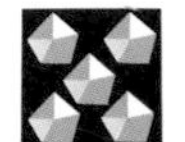

Smithsonite (or Bonamite or Dry-bone Ore)

Mohs' Hardness 4–5	Specific Gravity 4.3–4.5	Crystal Structure Hexagonal

Smithsonite was named after James Smithson (1765–1829), the founder of the world-famous Smithsonian Institution in Washington DC. It is used as a source of zinc and samples which have particularly striking bands will often be cut, polished, and used for ornamental purposes. Smithsonite occurs in voids and as stalactite-like layers in limestone cavities.

Smithsonite is white when pure, but will adopt any number of colors depending on the presence of additional elements or minerals: copper (usually malachite) will give a green to blue-green color; cadmium will give a bright yellow color; and cobalt a pink color to the crystals. Combinations of the above with the addition of small quantities of iron, manganese, magnesium, or lead can give brown or violet crystals.

Smithsonite sometimes occurs as crystals, but is often found in solid (sponge-like) form resembling dry bones—hence the name bonamite or dry-bone ore. Smithsonite is infusible, but soluble (with effervescence) in cold hydrochloric acid.

Range: Smithsonite deposits occur in Australia, Austria, the CIS, England, Greece, Namibia, Sardinia, Spain, Turkey, and the United States (Arkansas, Colorado, and New Mexico).

Variscite (Utahlite)

Mohs' Hardness 4–5	Specific Gravity 2.4–2.6	Crystal Structure Orthorhombic

Variscite, which is often confused with turquoise, is used as an ornamental stone. It is formed by the infusion of phosphatic waters into aluminous-rich rocks. It rarely occurs as crystals but more normally as nodules and masses.

Variscite is usually pale green, yellow-green, or green with a blue tint. Crystals tend to display no cleavage but break easily, giving concoidal fractures and very smooth surfaces. It is translucent with a vitreous to waxy luster, and the crystals are infusible but will discolor if heated.

Range: Nodules over 3½ feet in diameter have been found in Utah (hence the name). Quality stones are also found as nodules and masses in Austria, Bolivia, and the United States (Arkansas and Nevada).

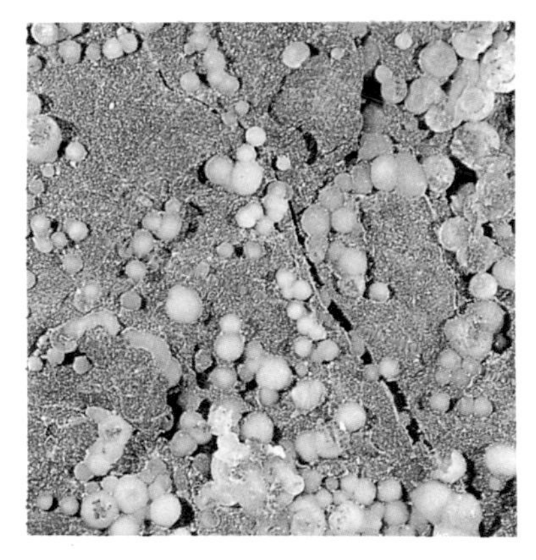

Fluorite (or Fluorspar) *halide group*

Mohs' Hardness 4	Specific Gravity 3.1–3.33	Crystal Structure Isometric

Fluorite is one of the mainstays of the modern chemical industry. It is used in the production of hydrofluoric acid, which is essential for petrochemical production, in plastics manufacturing, and oil-well stimulation.

Fluorite is usually associated with medium- and high-temperature hydrothermal veins, while the best-quality fluorite crystals have generally grown in cavities or been deposited in basins containing salt-rich waters.

The color of fluorite crystals varies enormously: from colorless and completely transparent when pure, to yellow, green, pink, purple, or violet, and blue or black when certain elements become tied up within the crystal structure.

Fluorite crystals will fluoresce (hence the name fluorspar) a bright violet or blue color. The crystals, which usually have an uneven or splotchy distribution of hues, are insoluble in water and most acids, with the exception of concentrated sulfuric acid. Crystal edges will fuse fairly easily, turning the flame brick red.

Range: Colorless crystals are found in Italy, pink crystals in Switzerland, green crystals in Norway, and violet crystals in England. Mineable reserves of fluorite are found in Canada, the CIS, England, Italy, Mexico, and the United States (Illinois and Kentucky).

Pyrrhotite

Mohs' Hardness $3\frac{1}{2}$–$4\frac{1}{2}$	Specific Gravity 4.6–4.7	Crystal Structure Hexagonal

Pyrrhotite, by itself, is only of use and interest to mineral collectors, but is more often than not associated with copper, iron, and sulfur. It is also a major source of cobalt, nickel, and platinum. Pyrrhotite is fairly common in mafic and ultramafic extrusive rocks, but also occurs in pegmatites, high-temperature hydrothermal veins, and occasionally as a sedimentary deposit.

Pyrrhotite crystals are usually tabular with striated faces. It fuses easily, changing from slightly magnetic to strongly magnetic, and gives off hydrogen sulfide fumes when the crystals are dissolved in hydrochloric acid. Pyrrhotite frequently occurs as a granular mass which is a reddish-brown color with a shiny metallic bronze luster.

Range: Substantial pyrrhotite deposits occur in Canada (Manitoba and Ontario). Quality crystals have been found in Brazil (Minas Gerais), Mexico, and the United States (Maine and New York State).

Rhodochrosite (or Inca-rose) *calcite group*

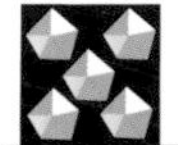

Mohs' Hardness $3^1/_2$–$4^1/_2$	Specific Gravity 3.3–3.7	Crystal Structure Hexagonal

Rhodochrosite is used as a source for manganese when it is available in large enough quantities. Sometimes banded masses of rhodochrosites are cut and used as ornamental stone or in large *objets d'art* where the markings can be displayed to their best advantage. Rhodochrosite crystals are usually found in hydrothermal veins and in sedimentary deposits. The term Inca-rose is derived from its formation as stalactites in abandoned mines which the Incas worked for silver.

The name rhodochrosite is derived from the Greek word *rhodon*, meaning pink, as crystals are normally a pink, faded pink or pinky-orange color. Chemical alteration can, and will, turn crystals brown or black. Crystals, which display a translucence and a vitreous to pearly luster, develop a dull oxidizing film of manganese when exposed to air. Rhodochrosite blackens gradually when heated.

Range: Rhodochrosite is quite a common mineral with important crystal deposits occurring in Argentina, the CIS, Mexico, Namibia, Romania, South Africa, Spain, and the United States.

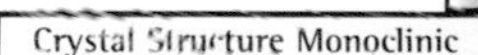

Azurite (or Chessylite)

Mohs' Hardness $3^1/_2$–4	Specific Gravity 3.7–3.9	Crystal Structure Monoclinic

Azurite was formerly used as a source of azure (blue) pigment. Due to its softness it is not very useful as an ornamental stone; it is, however, a copper ore of marginal importance. Azurite is a secondary copper mineral which occurs in sulfide deposits associated with carbonate rocks. It forms at lower temperatures than malachite, which often replaces it by ion exchange which occurs in an aqueous environment.

Azurite crystals tend to be azure-blue (hence the name) in color, but can also be dark blue with a green hint. Azurite crystals have a vitreous luster, are soluble in ammonia and effervesce in dilute acids. Crystals will fuse easily, first turning black as they give up their water. Powdered azurite will, with time, turn greenish as it alters to malachite.

Range: The name chessylite is derived from Chessy, the name of the town in France where spherical aggregates are found. Azurite deposits also occur in Australia, Chile, Greece, Iran, Mexico, and Namibia.

Chalcopyrite (or Copper Pyrite)

Mohs' Hardness 3½–4	Specific Gravity 4.1–4.3	Crystal Structure Tetragonal

Chalcopyrite ore accounts for almost 80 percent of the world's copper metal. It is fairly common and widespread in its distribution, and often yields gold and silver as a by-product. Chalcopyrite occurs as a sulfide in high-temperature hydrothermal veins.

Chalcopyrite crystals are a brass color, but over time tarnish to a range of colors (often with an iridescent film) depending on the environment and the chemical composition of the original crystal and the host rock. They are opaque with a metallic or sub-metallic luster and will burn, coloring the flame green, and giving off a highly unpleasant gas.

Range: Substantial chalcopyrite deposits occur in Canada, Chile, the CIS (the Urals), Cyprus, the United States (Arizona, Montana, and Utah), Zaire, and Zambia. The Bingham Canyon copper mine in Utah is reputed to be the largest man-made hole in the world. The reclaiming of the massive open pit—due to commence once extraction had ceased—was going to be a major undertaking; that will now no longer be necessary as the area has been granted "monument" status and attracts thousands of visitors a year.

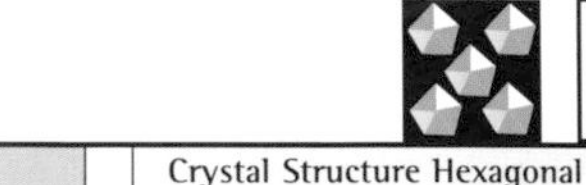

Dolomite

Mohs' Hardness 3½–4	Specific Gravity 2.85–2.95	Crystal Structure Hexagonal

Dolomite is used as a structural and ornamental stone and for producing special cements. It is used in the manufacture of magnesia for refractories, as a metallurgical flux for the iron and steel industry, as well as being used as a source of magnesium. Dolomite occurs most commonly as a sedimentary deposit believed to be caused by calcite alteration. It is typically deposited in low-temperature hydrothermal veins, or is formed due to the partial metamorphism of dolomitic limestone into marble.

Dolomite is usually a collection of small, colorless, white, pale gray, pinkish, or yellow-tinted crystals, frequently found in a saddle-shaped aggregate of slightly curved crystals. Dolomite fragments will dissolve with difficulty in hydrochloric acid and are not fusible. Dolomite resembles calcite in appearance and form, but calcite will effervesce and dissolve readily in cold hydrochloric acid.

Range: Thick beds of sedimentary dolomite occur throughout the world. The finest quality crystals come from Brazil (Bahia), Canada (Quebec), Italy, Switzerland, and the lead-zinc mining area of the United States (Missouri).

Malachite

Mohs' Hardness 3½–4 | Specific Gravity 3.75–4.00 | Crystal Structure Monoclinic

Malachite is used as an ornamental stone which—when cut into slabs and polished—is often made into boxes, small tables, and *objets d'art* where its distinctive banding can be shown to its full advantage. Malachite, despite its exotic and distinctive markings, has rather humble origins; it is a secondary ore occurring in the upper levels of copper deposits, where it has been altered by the action of carbonated water.

It occurs in microcrystalline masses, usually as nodules with radiating bands. Malachite ranges in color from weak green to emerald green and from deep, dark green to blackish-green. It is fragile, has good cleavage, and holds a silky polish well. It fuses fairly easily, coloring the flame green due to its copper content. It turns dark green and then black as it gives up its water, finally leaving a blob of metallic copper. Malachite crystals are rarely found and are usually twinned.

Range: The most important sources of malachite in the form of banded masses are in the CIS (Siberia and the Urals), Zaire, Zambia, and Zimbabwe. Less important sources occur in Australia, Chile, France, Namibia, and the United States (Arizona and New Mexico).

Sphalerite (or Zinc Blend)

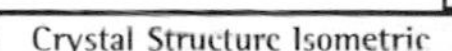

Mohs' Hardness 3½–4 | Specific Gravity 3.9–4.2 | Crystal Structure Isometric

The name sphalerite is derived from the Greek word meaning "treacherous," possibly due to the variations which commonly occur in its appearance. Sphalerite is a major source of zinc (used in alloying and galvanizing); it provides cadmium, gallium, and indium as by-products. Sphalerite is usually associated with hydrothermal activity, and it occurs frequently with barite, chalcopyrite, fluorite, and galena. Sphalerite sometimes occurs in sedimentary deposits or in low- and medium-temperature deposits where replacement crystal substitution has taken place.

Sphalerite varies in color from yellowish-brown to reddish-brown (when pure), to blackish-brown when iron is present. It can also be green, pink, pinkish-orange (called "honeyblend"), red (sometimes incorrectly called "ruby-red"), or colorless. Iron-rich sphalerites have a submetallic luster while other variations have a resinous luster. It is soluble in hydrochloric acid and infusible if pure, but will become more fusible as the iron content increases.

Range: Gem-quality sphalerite is found in Mexico and Spain. Less important deposits occur in England, Sweden, the U.S., and the former Yugoslavia.

Chalcocite (or Redruthite or Copper Glance)

Mohs' Hardness $2^1/_2$–3	Specific Gravity 5.5–5.8	Crystal Structure Orthorhombic

The name redruthite comes from the name of the town Redruth in Cornwall (England) which was the main important source of the world's copper and tin industry for much of the last century. Chalcocite is an important source of copper metal and is associated with hydrothermal copper-sulfide deposits as a secondary-enriched zone.

Chalcocite rarely occurs as crystals, but when it does, they are soft, malleable, pseudo-hexagonal, and striated. Chalcocite more typically occurs as a granular, dull gray, aggregate, whose surface has altered to green or black. It is fusible and dissolves easily in nitric acid.

Range: Quality chalcocite crystals have been found in Chile, England, Mexico, Namibia, South Africa, Spain, and the United States (Connecticut and Montana).

Copper (or Native Copper)

Mohs' Hardness $2^1/_2$–3	Specific Gravity 8.93–9.00	Crystal Structure Isometric

After iron, copper is probably the most important metal in man's history. Used as an alloy for centuries, it is now the most important metal in the field of electrical engineering. Copper is typically formed by reduction in oxidization zones of sulfide deposits or, in mined-out sites where copper-sulfated water will precipitate onto iron objects, or replace organic fibers, such as wood.

Copper is copper-red when freshly broken, but tarnishes rapidly to a dull, dusty looking brown color. Crystals are soft, have a metallic luster, and fracture in an irregular manner. Copper rarely occurs as crystals; it is found more frequently in compact masses or branch-like shapes. Copper is very malleable, and dissolves easily in nitric acid, staining the resulting solution an azure-blue color.

Range: The finest crystals of copper come from the United States (Michigan). Less spectacular examples come from deposits in Chile, the CIS, Germany, Mexico, Spain, Sweden, and Zambia.

Galena

Mohs' Hardness $2\frac{1}{2}$–3	Specific Gravity 7.2–7.6	Crystal Structure Isometric

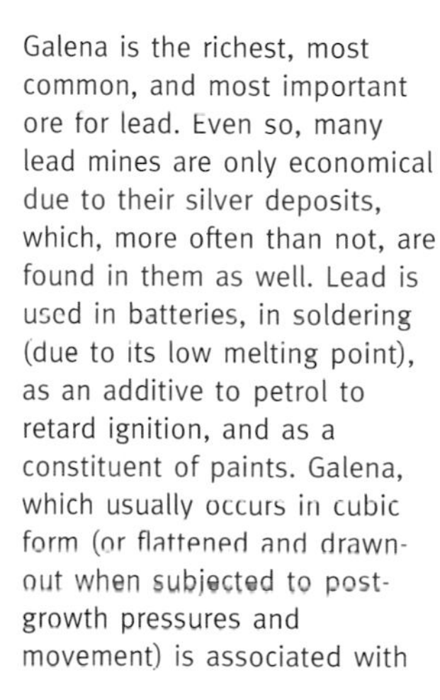

Galena is the richest, most common, and most important ore for lead. Even so, many lead mines are only economical due to their silver deposits, which, more often than not, are found in them as well. Lead is used in batteries, in soldering (due to its low melting point), as an additive to petrol to retard ignition, and as a constituent of paints. Galena, which usually occurs in cubic form (or flattened and drawn-out when subjected to post-growth pressures and movement) is associated with calcite veins contained in limestone or dolomitic masses.

Crystals are dull gray and cubic in shape. They have a warm feeling, will leave a gray mark on the hands if rubbed, and are heavy. They are soluble in heated hydrochloric acid, and give off an unforgettable "bad egg" smell.

Range: Important lead deposits occur in Australia, England, the United States (Idaho and Missouri), and western Germany.

Silver

Mohs' Hardness $2\frac{1}{2}$–3	Specific Gravity 10.5 pure	Crystal Structure Isometric

Known to the Incas as the "tears from the moon," silver is a precious metal which can occur in crystal form. Silver and its derivatives are used in industry as a chemical catalyst; in electronics, because of its high conductivity; and in dentistry, due to its inertness. It is also used in jewelry. By far the largest use for silver is in the field of photography, because of the sensitivity of silver bromide to light.

Silver crystals are rarely found, but when they do occur, they are associated with hydrothermal veins; there, they are wire-like, soft, malleable, and silver-white when freshly broken. They tarnish rapidly to brown, dull-gray, or black. Silver, however, more generally occurs in irregular masses as silver associated with gold and/or copper.

Range: The finest wire-like crystals come from Norway; they are also found in eastern Germany and Mexico. Important silver deposits are found in Australia, Canada, Chile, and the United States (Colorado, Nevada, and South Dakota).

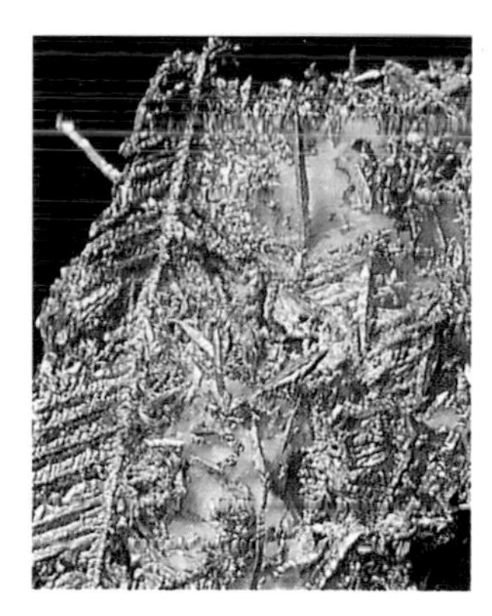

Brucite

Mohs' Hardness $2^1/_2$	Specific Gravity 2.3–2.5	Crystal Structure Hexagonal

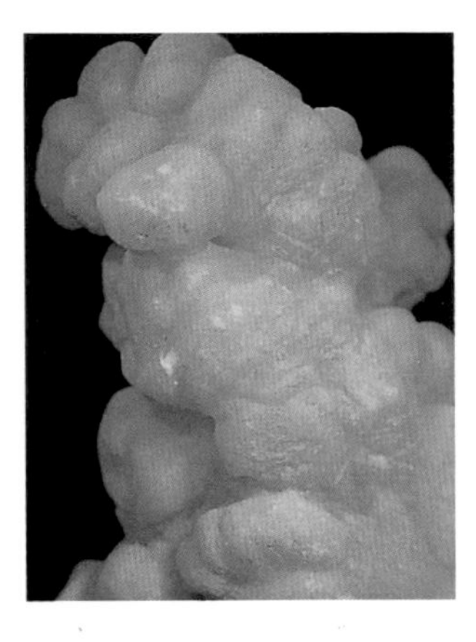

Brucite is widely used as a refractory material in the extraction of magnesia and as a source of metallic magnesium and its salts. It is a metamorphic mineral found in low-temperature serpentized rocks or in contact metamorphized dolomites. It is also found as an alteration product in the final stages of a metamorphic incident.

Brucite commonly occurs as scaly, plate-like, or highly fibrous aggregates. It is transparent or translucent, being either colorless, white, yellowish, or occasionally pink. It is infusible but dissolves readily in cold diluted acid with no effervescence. It parts easily along micaceous cleavage planes, which then display a dullish-pearly luster. Brucite is often an associated mineral of asbestos.

Range: Brucite occurs in Canada, the CIS, Scotland, and the United States (Nevada and Pennsylvania).

Gold

Mohs' Hardness $2^1/_2$	Specific Gravity 19.3 pure	Crystal Structure Isometric

Known to the Incas as the "sweat of the sun," gold is a rare precious metal which can occur in crystal form. It has been fashioned into ornaments and jewelry since the dawn of time. During the last millennium gold has become established as a way of storing wealth, and as a hedge against the economic effects of natural disaster, war, and man's folly. In more recent times, gold's ability to conduct heat and electricity without ever tarnishing has led to it being used widely in technology and industry.

Though primarily of hydrothermal origin, large concentrations of gold are formed by the erosion and re-deposition of gold-bearing lavas. Gold crystals are golden-yellow when pure ("yellow gold") tending to silvery-yellow when alloyed with silver ("white gold"), and reddish-orange when mixed with copper ("red gold"). The distinguishing features of gold are its weight and the fact that it will never tarnish or rust.

Range: Of the major gold-producing areas in the world, the Witwatersrand district in the Transvaal (South Africa) and the Carlin Trend in Nevada (United States) are the most famous. Australia, Brazil, Canada, and Papua New Guinea also contain major gold deposits.

Halite (or Common Salt or Rock Salt)

Mohs' Hardness $2\frac{1}{2}$	Specific Gravity 2.1–2.2	Crystal Structure Isometric

Halite has been used for thousands of years as a means of preserving meat. It is used today for this purpose and in food preparation in general. Its most important market is in the chemical industry for manufacturing soda, sodium, and hydrochloric acid.

Halite is predominantly an evaporate, often aided by precipitation in enclosed seas. Such deposits are overlaid often by bands of clay and shale; however, being plastic in nature and having a low density, halite tends to flow upwards. The resulting "salt domes" are often worked for their halite or sulfur, or for the hydrocarbon traps which often occur at the edges of the salt dome and the host rock.

Halite crystals are commonly colorless, white, yellow, or red. They have a vitreous luster, are brittle, have a pleasant salty taste, and over a period of time will dissolve in the water that they have absorbed from the atmosphere.

Range: Large halite deposits occur in Austria, the CIS (Siberia), the former Czechoslovakia, England,

France, Germany, Poland, and the United States (Louisiana and Texas).

Muscovite (or Common Mica or White Mica)

Mohs' Hardness 2–3	Specific Gravity 2.77–2.88	Crystal Structure Monoclinic

Muscovite is an important industrial material as it is a superb electrical and heat insulator and is used for this purpose either in small sheets, or is powdered and re-formed with cement and plastic. It is used in the oil industry as a plugging material when fluids are being lost to porous or cavernous formations; as a filler in paper and rubber manufacture; and as an addition to badly drained soils.

Muscovite commonly occurs in silica- and aluminum-rich igneous or metamorphic environments. It is extremely common in sand and other sedimental materials—which are formed as a result of the disintegration and weathering of igneous and quasi-igneous rocks—where it has crystallized from gases.

Muscovite is characterized by its perfect cleavage and the resulting flexible flakes. Crystals are commonly pseudo-hexagonal, forming "mica blocks" within a matrix of sheets displaying no apparent structure. It can vary in color from colorless to black and can be also white, yellow, reddish, or brown.

Range: Huge single crystals (measuring eight yards across) have been found in Brazil, Canada (Ontario), India, and the United States (New Hampshire and South Dakota).

Vermiculite

Mohs' Hardness 2–3	Specific Gravity 2.4–2.7	Crystal Structure Monoclinic

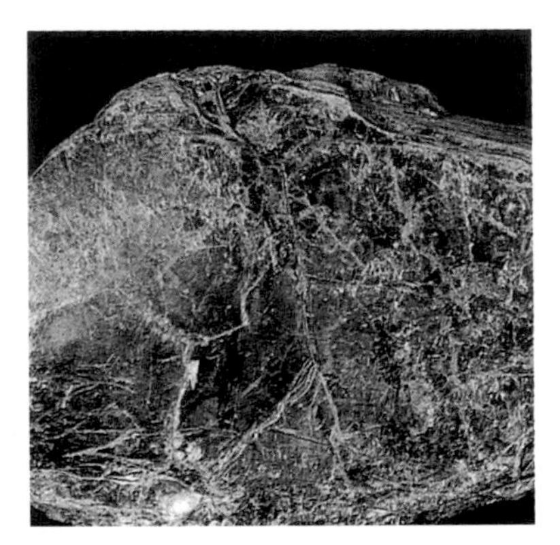

Vermiculite is a clay-form, which when heated rapidly expands dramatically. Expanded vermiculite is an excellent thermal and acoustic insulator and is used for this purpose in the building industry. In recent years demand for it has increased due to the gap left in the market when the health problems associated with asbestos were recognized. Vermiculite is also used as a filler in the paper, rubber, and plastics industry, and as a packaging medium. It has recently gained much favor as an additive to heavy clay and badly irrigated soils because it can increase water percolation. Vermiculite is caused by the hydrothermal alteration of biotite and phlogopite.

Vermiculite is a plate-like, golden, or honey-colored crystal which frequently occurs as a scaly aggregate. It has a pearly vitreous luster, is slightly soluble in acids and expands by up to as much as 20-fold when heated beyond 572°F.

Range: The largest vermiculite deposit occurs in South Africa with other significant masses located in Argentina, Australia, Canada, and the United States (Massachusetts, Montana, and North Carolina).

Argentite (or Silver Glance)

Mohs' Hardness 2–2½	Specific Gravity 7.2–7.4	Crystal Structure Isometric

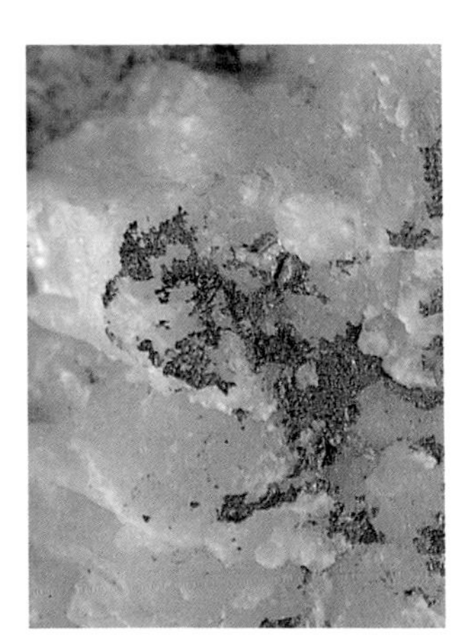

Argentite is an important silver ore. It frequently occurs with galena from which it is separated by melting the ore and skimming the silver off the top where it tends to float. Silver has been used for centuries as a metal in jewelry and for plating purposes.

Argentite is frequently found in low-temperature hydrothermal veins with other silver minerals, usually as a massive agglomeration, either in groups of branching crystals or in distorted form. It is soft and heavy, and tarnishes quickly to a ghostly, gray-green, dull color. Argentite is metallic and shiny when freshly cut or broken. It fuses easily into a silvery blob.

Range: Argentite crystals have been found in Bolivia, the former Czechoslovakia, Mexico, and Norway. As a mineral deposit, argentite is found in Australia, Canada, Chile, Peru, and the United States (Colorado and Nevada).

Bismuth (Native Bismuth)

Mohs' Hardness 2–2½	Specific Gravity 9.7–9.83	Crystal Structure Hexagonal

Bismuth is the main source of bismuth metal which is used in alloys with a low melting point, in lubrication additives, and in medicines and cosmetics. It is a fairly uncommon mineral which is found in hydrothermal veins where it is often associated with cobalt, nickel, silver, and tin ores.

Bismuth crystals are rare but, when they do occur, are usually imperfectly formed. Bismuth is very heavy, opaque with a bronze-colored metallic luster, and fuses easily at a low temperature to produce a metallic blob. Bismuth dissolves easily in nitric acid, which will then yield a white precipitate if diluted with water. Bismuth usually occurs in massive form or in a foliated aggregate.

Range: Substantial bismuth deposits occur in Bolivia. It can also be found as an associate mineral with lead, cobalt, and silver in Canada, eastern Germany, and Norway.

Borax

Mohs' Hardness 2–2½	Specific Gravity 1.70–1.74	Crystal Structure Monoclinic

Borax is used as an industrial cleaning agent, in the manufacture of high-temperature glass (such as pyrex), and as a flux in soldering, brazing, and welding. It is also used as a source of boron, which has the ability to absorb neutrons and is used in the control rods employed in nuclear reactors. Borax is an evaporate which has formed due to the evaporation of ponded-in saline lakes and seas.

Borax normally occurs as soft, prismatic, stubby crystals, which range in color from colorless to white and from whitish-blue to yellowish-white. Colorless crystals tend to lose their water and assume a dusty white appearance not unlike square marshmallows. Borax is water soluble and has a slightly sweet, alkaline taste.

Range: In ancient times, borax was brought to Europe from the salt lakes of Tibet. Today, the main deposit is in the United States (Death Valley, California) with smaller crystal deposits being found in Argentina and Turkey.

Cinnabar

Mohs' Hardness 2–2½	Specific Gravity 8.0–8.2	Crystal Structure Hexagonal

Cinnabar is the most important source of mercury metal, and was used in the past as the mineral pigment known as vermillion (or vermilion).

Cinnabar crystals are rare, but mostly occur as a coarse, granular, or compact aggregate which is found in masses and impregnations, in lavas, and near hot spring deposits. It also occasionally occurs in sedimentary deposits which have acted as a replacement medium from nearby igneous rocks.

Cinnabar is opaque, with a dull, pink, washed-out, earthy-red color. It is soft, heavy, and deposits droplets of mercury on cold surfaces after being heated. It is insoluble in acids, but can be attacked by aqua regis and chlorine gas. Cinnabar crystals fracture unevenly, are brittle, and have a tendency to splinter.

Range: For industrial purposes, the most important cinnabar deposits occur in Italy, Spain, and the former Yugoslavia. Smaller deposits occur in Algeria, China, the CIS, Peru, and the United States (Arkansas, California, Texas, and Utah).

Gypsum

Mohs' Hardness 2	Specific Gravity 2.35	Crystal Structure Monoclinic

Gypsum is the most common sulfate mineral. Massive stratified gypsum deposits are worked in the Paris basin.

Much of this is used in the manufacture of plaster of Paris, from which the name is derived. This remains the principal use for gypsum. In Europe powdered limestone is used as a filter to reduce sulfur gas emissions from coal-burning power-stations. This limestone is chemically changed into a type of "dirty gypsum" which is unsuitable for fine plaster-work where whiteness is required, but it is still used in other forms of building work. Gypsum is also used as a retarder in cement, as a fertilizer, and as a flux in glass fabrication.

Gypsum, which can be colorless, white, yellow, gray, or brown, commonly occurs as elongated flattish crystals with a silky luster. Crystals are frequently bent, and large crystals are not uncommon.

Range: Splendid gypsum crystals occur in Chile, Mexico, Sicily, and the United States (Utah), while extensive deposits are mined in the CIS, England, France (Paris), and the United States.

Stibnite

Mohs' Hardness 2	Specific Gravity 4.6–4.7	Crystal Structure Orthorhombic

Stibnite is the main ore for antimony which is alloyed with tin and copper to produce an anti-friction alloy. It is also alloyed with the lead used in storage battery plates. Antimony is used in a number of ways—the manufacture of matches, as a color generator in fireworks, and as an industrial pigment. The slats derived from stibnite are used in the vulcanizing of rubber, in medicine, and in glass fabrication. Stibnite is a low-temperature mineral that often occurs in hydrothermal veins.

Stibnite is soft and steely-gray, commonly occurring as either spines radiating out from a base or as striated columns. They are soft, often bent, and have a brilliant metallic luster. Thin splinters will fuse in a match flame, whilst powdered stibnite is soluble in concentrated hydrochloric acid. A few drops of this solution will precipitate bright orange in potassium iodine.

Range: Superb crystals, sometimes over a foot in length, have been found in China, Japan, and Romania.

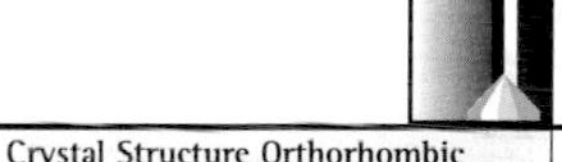

Sulfur (or Native Sulfur)

Mohs' Hardness $1\frac{1}{2}$–$2\frac{1}{2}$	Specific Gravity 2.00–2.10	Crystal Structure Orthorhombic

Sulfur is mostly used for the manufacture of sulfuric acid, which is then used to make fertilizers. It is also used for vulcanizing rubber, and in the manufacture of explosives and insecticide. Sometimes referred to as the "Devil's milk" due to its distinctive smell, most of the sulfur produced today is as a by-product of hydrocarbon production and processing.

Large sulfur lenses do, however, occur at the top of salt domes in the Gulf of Mexico, off the coast of the United States. These deposits, which are often discovered when drilling for oil, are "mined" using the Frasch process. Superheated water is pumped down a borehole into the formation, melting—but not dissolving—the sulfur. The "melt" is then aerated with hot air bubbles (to keep it fluid) and lifted to the surface via a pipe set in the same borehole.

Pure sulfur is yellow, but is more likely to be beige, dull brown, or black (especially in Iran). The darker the color the higher the entrapped hydrocarbon impurities. Sulfur crystals are translucent and burn with a blue flame.

Range: The most prolific area of sulfur production is the United States (Louisiana and Texas) with sublimate deposits occurring in Chile, Japan, and Sicily.

Bauxite

Mohs' Hardness 1–3 | Specific Gravity 2.00–2.7 | Crystal Structure Aggregate of various crystal types

Bauxite is one of the world's most commonly occurring ores. Its extraction is often dependent on the availability of cheap transport and/or cheap electricity. It requires cheap transport since bauxite mining is a victim of the "economy of scales" syndrome; cheap electricity is required since this is needed to refine bauxite into aluminum. Bauxite is a secondary deposit: it is what remains after aluminum-bearing rocks have been weathered in tropical and subtropical climates.

Bauxite is commonly red-brown to dull brown, but can also be white, yellowish, or gray in color. It is dull, finely grained, often earthy in texture and appearance. Generally, it is not a very interesting or prepossessing mineral. Nonetheless, it is the world's main source of aluminum.

Range: Important bauxite deposits occur in the CIS, Ghana, Hungary, Indonesia, Jamaica, and Surinam.

Graphite

Mohs' Hardness 1–2 | Specific Gravity 2.09–2.23 | Crystal Structure Hexagonal

Graphite is used in lead pencils, and is mixed with mineral oils or grease to form a high-temperature lubricant. It is also used in the form of electrodes, as brushes in electrical motors, and in protective paints. Graphite usually occurs in metamorphic rocks as the final stage in the carbonization of organic substances. Graphite is occasionally found as thin hexagonal crystals with triangular basal markings, but it occurs more frequently as foliated masses or thin sheets.

Graphite is steely-gray to black, or opaque with a sub-metallic dull luster. It is greasy to the touch and very soft. Graphite is used as one of the comparison materials in Mohs' hardness test kits. It will form a gray-black streak on paper. It displays perfect basal cleavage which results in thin platelets. It is insoluble in acid and will only melt in extremely high temperatures.

Range: Significant deposits of graphite are found in the CIS, the former Czechoslovakia, Madagascar, Mexico, South Korea, and Sri Lanka.

Molybdenite

Mohs' Hardness 1–1½	Specific Gravity 4.6–5.0	Crystal Structure Hexagonal

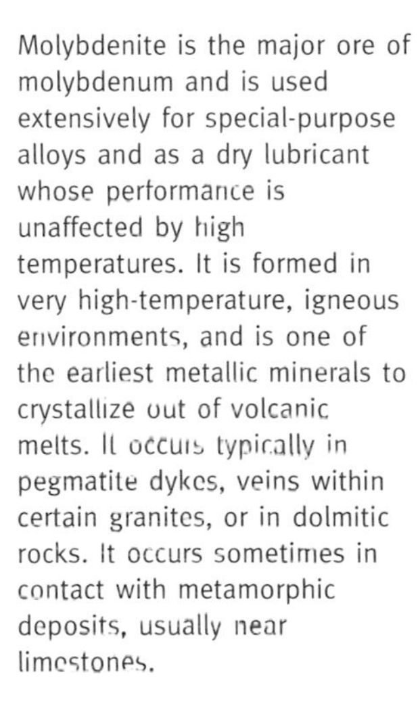

Molybdenite is the major ore of molybdenum and is used extensively for special-purpose alloys and as a dry lubricant whose performance is unaffected by high temperatures. It is formed in very high-temperature, igneous environments, and is one of the earliest metallic minerals to crystallize out of volcanic melts. It occurs typically in pegmatite dykes, veins within certain granites, or in dolmitic rocks. It occurs sometimes in contact with metamorphic deposits, usually near limestones.

Molybdenite crystals are rare, but when they do occur are lead-gray or bluish-gray in color and not dissimilar to graphite or galena in appearance. Molybdenite more commonly occurs as foliated or scaly aggregates. These are easily separated along perfect cleavage planes into flexible platelets. Molybdenite is greasy to the touch, opaque, and with a luster which is sometimes metallic in character. It can also appear to be tired and dull.

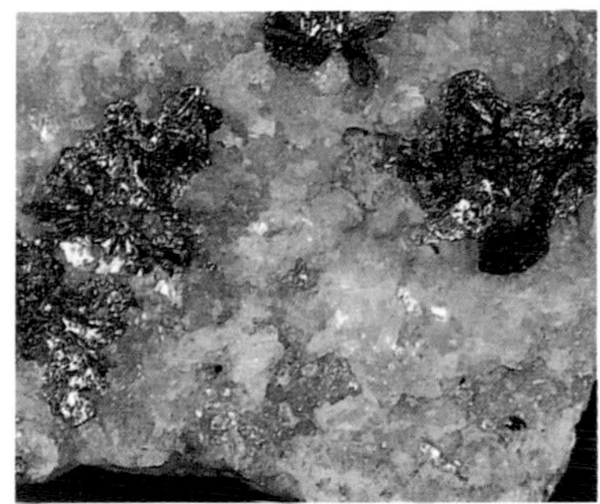

Range: Molybdenite occurs in Canada, Chile, China, the CIS, Norway, and the United States (Colorado).

Montmorillonite (or Bentonite)

Mohs' Hardness 1	Specific Gravity 1.2–2.7	Crystal Structure Monoclinic

Montmorillonite is the chief constituent of the clay mineral bentonite. Montmorillonite absorbs water—especially water that has a high alkaline value—and swells accordingly. A liquified clay like this is used in the oil drilling and construction industries due to its ability to suspend solids. It can also seal the pores of formations which normally accept fluids. Montmorillonite will become semi-solid when left undisturbed, but will become pumpable again when agitated (this is called being thixotropic).

It is also used as a purifying medium and as a filler in the manufacture of paper and rubber. It is formed either in a hydrothermal environment where volcanic ash has been altered, or in a sedimentary tropical environment where feldspars have been altered. It is a microcrystalline material which is white, gray, or beige. It forms earthy masses which are greasy to the touch and crumble easily.

Range: Montmorillonite is found in large masses in Montmorillon in France (hence the name), Germany, Japan, and the United States.

Talc (Soapstone)

Mohs' Hardness 1 | Specific Gravity 2.58–2.83 | Crystal Structure Monoclinic

Talc is usually powdered and used as a filler in paper, rubber, and paints. It is also used in the textile and cosmetics industries. Slabs of talc are used to acid-proof laboratory surfaces, while larger pieces are often fashioned into simple statues. Talc is a secondary mineral formed at the metamorphic alteration of any one of a number of magnesium silicates.

Talc never occurs as visible, individually formed crystals, but as a white, gray, or greenish, pearly, foliated compact mass. It is greasy to the touch (hence the name soapstone) and is used as the lowest of the comparison materials in Mohs' hardness scale. Talc is insoluble in acid and almost impossible to fuse.

Range: The largest talc mine in the world is in the Pyrenees mountains (France), where high-quality talc is still selectively mined by hand in what is a primitive but highly profitable operation. Due to the altitude and weather, the open-cast mine is operated for only six months a year. Smaller deposits occur in Australia, Austria, Canada, India, Korea, and South Africa.

Mercury

Mohs' Hardness zero—at room temp. | Specific Gravity 13.6 | Crystal Structure Hexagonal—below −38°F

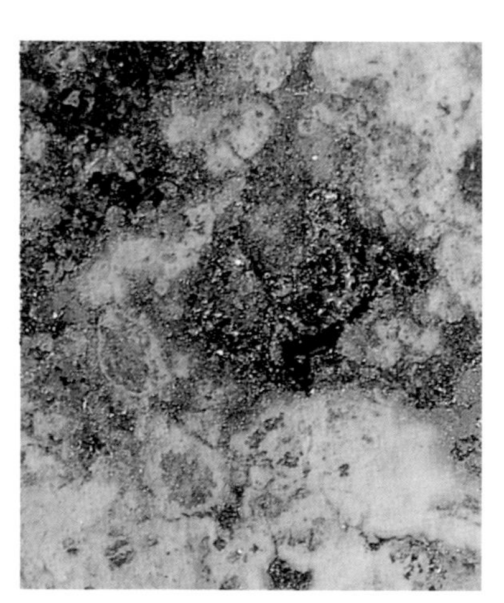

Mercury is used in mineral processing for recovering gold and silver. It is also used in the manufacture of explosives, batteries, and electrical rectifiers. During the last century, mercury was used for curing beaver furs which were made into hats. The vapors given off were eventually found to be poisonous and to cause insanity. Mercury usually occurs in association with cinnabar in areas of volcanic or geyser activity.

Mercury is the only metal which is liquid at room temperature. It usually occurs in nature as tiny red beads which appear to be weeping out of bright red cinnabar. Its liquid nature, its association with cinnabar, and its weight are distinctive features of the metal. Mercury will dissolve in nitric acid.

Range: The most important mercury deposits are found in Italy, Spain, and the former Yugoslavia.

Glossary

Acicular Crystals which display a very long, needle-like habit. Often radiating from a base.
Aggregate An assembled mass of more than one crystal type and usually more than one mineral species.
Alluvial Material which has been transported by water and accumulates to form a rock mass.
Aqua Regis A mixture of one part of hydrochloric acid to one part of nitric acid. Capable of dissolving gold.
Asterism An optical effect which appears as a star-shape within a crystal. Caused by minute secondary crystal inclusions.
Authigenic A mineral formed within a sedimentary environment and not one which has been introduced.
Birefringent An optical phenomenon which causes any image viewed through such a crystal to have a double appearance.
Carat A unit of weight used for gemstones. One carat is equal to 3 grains (200 mg).
Cleavage The plane along which a crystal will naturally break. This may not necessarily be a well-defined plane.
Columnar A crystal shape which resembles a column; a regular elongated prism.
Conchoidal A crystal that breaks with concentric cavities, eg quartz – subconchoidal-indistinct conchoidal.
Concretion A concentration of mineral growth around a nucleus within a sedimentary environment.
Contact The transmission of temperature from an igneous rock, without a pressure change; often leads to a metamorphic change.
Dendrital The skeleton form of a crystal usually found on fracture surfaces; capable of absorbing matter.
Density Ratio of the weight of an object to its volume. Frequently referred to as the specific gravity.
Deposit An accumulation of minerals in sufficient quantity and quality so as to be economically extractable.
Dike An igneous intrusion of great length but limited thickness which often fills a vein or fracture plane.
Ductile Bends easily, usually without any permanent damage (like warm licorice sticks).
Eruptive Being ejected from below the earth's surface; lava, dust, ashes, gases, and vapors.
Evaporite A mineral of chemical origin which owes its formation to the evaporation of an aqueous solution.
Extrusive When molten rocks or lava flow out of the earth's surface.
Fluorescence The temporary emission of light waves to give colors not normally seen.
Foliated The crystal habit that resembles foliage (hence the name). Minerals are usually flaky in character.
Fracture To break unevenly. This occurs in crystals which do not have a clearly defined cleavage.
Gangue That part of a mineral which is of no value but which must still be extracted.
Gel A semi-solid solution, usually heavily charged with elements; often solidifies to form colored minerals.
Geode Spherical cavity; much favored by crystals as a location to grow in.
Greisen Igneous rocks which have been altered by fluids, rich in volatile elements.
Habit The characteristic shape of a crystal; the form in which it most frequently occurs.
Hardness The resistance to wear. Often measured by means of the Mohs' scratch test scale.
Hydrothermal The process by which heavily charged aqueous solutions transport and form minerals.
Igneous A rock which is formed by the solidification and crystallization of molten rocks or magma.
Inclusion The entrapment within a growing crystal of a gas, a liquid, or a solid (often as a secondary crystal growth).
Leaching Zone The area in the lower part of a mineral deposit, where chemical compounds are leached and then redeposited.
Luster The light reflected off the surface of the crystal.
Mafic The characteristic of a mineral which refers to its predominance of ferromagnesium minerals.
Magmatic The formation of minerals from molten silica-rich rocks, contained within the earth's surface.
Malleable Soft material (usually refers to metals) which can be formed into shape without the need to be beaten.
Metamorphism The transformation of a rock from one state to another; caused by the effects of a nearby heat and pressure source.
Mineral A naturally formed homogeneous solid which possesses a clear chemical composition.
Oxidation The chemical process by which oxygen is added to a compound or process. Opposite to reduction.
Pegmatites Igneous intrusions caused when residual liquids cool from magmas.
Piezoelectric Ability of a crystal to emit an electrical charge if compressed, or vice-versa.
Pipe A volcanic structure, usually tube-like, through which magmatic material is forced upwards.
Pleochroism The ability of a mineral to reflect and absorb different colors, depending on their direction of orientation.
Reduction The chemical process by which oxygen is removed from a compound or process. The opposite chemical process to oxidation.
Schists Metamorphic rocks which contain mineral depositions (usually mica-like) in parallel or subparallel veins.
Synthetic A precious mineral reproduced in a laboratory, with the same characteristics as that of a natural stone.

Ancient Uses and Healing Powers

agate

Agates are believed to be a stabilizing stone, and as such can be used to calm a troubled mind or body. They are an ideal stone to use as a "worry bead" as they reputedly have the ability to dispel anxiety. They can also be used to counter the effect of other stones which someone else may be wearing to your disadvantage.

amethyst

The amethyst crystal is thought to have the power to transmit calm. It is also believed to uplift the mind and to improve general well-being, as well as alleviating insomnia. It is reputed to have a powerful healing influence and generates spiritual awakening. It can also be used as an aid to chastity as it is believed to calm strong emotions.

aquamarine

Aquamarines are renowned for their tranquil qualities and their ability to bring peace and calm to the most troubled of minds. The crystals have a cooling power and as such can be used to calm a fever or relax an over-active mind. Aquamarines are thought to cure motion sickness as well as cleanse glands, relieve water retention and reduce nervous disorders. Aquamarine is the gemstone of the Gemini star sign.

citrine

Citrine is a sun stone; it is believed that it enables one to develop clear lines of thought, direct arguments, and precise communication skills. Citrine crystals are reputed to help the body rid itself of physical and emotional problems.

diamond

The diamond is considered the stone of the mind because of its hardness. It can be used to stimulate clear thoughts or prevent dreaming. The legendary brilliance of the stone is reputed to form a barrier against negative thoughts. Eros, the Greek god of love, tipped his arrows with diamonds to ensure that those whose hearts they struck would fall in love with each other. The diamond is the gemstone of the Libra star sign.

emerald

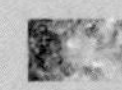

The emerald is the gem of love and is thought to aid fertility, growth, honesty, and self-fulfilment. The crystal can reputedly help one solve problems by stimulating the brain and memory. In the past emeralds have been used for curing skin disorders, while nowadays they have become accepted as an alternative to pharmaceutical antiseptics. The emerald is the gemstone associated with the Taurus star sign.

hematite

Hematite (or bloodstone), as may be expected, is associated with blood disorders and purification. It is reported to stimulate courageous behavior, both physical and emotional, and to restore the balance to highly charged situations. Women may find that the bloodstone is useful for regulating irregular menstrual flow.

jade

Jade, as one would expect from the sea-green color with which it is associated, is reputed to have a calming power, and as such is useful to those who suffer from any form of anxiety, tension, or stress-related ailment. It is believed to establish the foundations of true friendship as well as giving one the confidence to express true love. The ancient Chinese believed that jade would guarantee a long and prosperous life.

kunzite

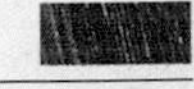

Kunzite crystals are considered essentially feminine in shape and purpose. They reputedly help rejuvenate the skin and are good for the heart. They are believed to stimulate self-praise and fulfilment and are reputed to regulate the menstrual cycle.

lapis lazuli

Lapis lazuli is often considered a suitable stone for starting one's spiritual birth. It is also believed to act as a medium through which positive psychic influences are transmitted. With a history of use going back over 7,000 years, its powers to develop inner strength and confidence are legendary.

malachite

Malachite is thought to be a calming stone; its ripple-like markings seem to radiate out from nowhere. It is reputed to stimulate passive resistance and inner awareness. In ancient times it was thought to have the power to strengthen teeth, aid those with poor eyesight, and warm a cold heart.

moonstone

The misty silver color of the moonstone is said to brighten at the start of every new moon. When passed between lovers, the moonstone is reported to arouse a feeling of warmth and friendship which many physical relationships fail to develop. The moonstone is a sensitive stone and is very susceptible to mood changes.

opal

Due to the ease with which opals will give up or take up water, they are often considered to be a living stone which can move forward in the future. It is not surprising that some people believe that such journeys age the opal beyond its limitations and cause ageing, which leads to it cracking. Opals are a feminine stone, once thought to guard against gray hair in blonde women. It was also believed to make childbirth less painful. Due to their cantankerous nature, opals should be kept away from other stones.

ruby

The ruby is often considered to be the prince among crystals as it is said to contain the heritage of humanity, the gift of humility, and the beauty of a pure spirit. Rubies help the body to survive in times of peril and are often associated with authority and natural leadership. Over the centuries, the ruby crystal has become a symbol for everlasting love and loyalty. In ancient times it was believed that the stone contained an inner heat which had the power to boil water. Rubies are associated with the Leo star sign.

sapphire

The sapphire is known as the jewel of truth and wisdom and is associated with heavenly inspiration, devotion, and spiritual control. The crystal is also reputed to be capable of controlling desire and passion. Some cultures believe that the three points contained within the star sapphire represent destiny, faith, and hope; consequently these stones are considered to be particularly useful for transforming wishes into reality.

topaz

Topaz means "fire" in Hindu, and in living up to its name brings light to life. The stone is believed to relieve stress; a crystal placed under the bed at night is reputed to have revitalizing properties, stimulating refreshing dreams. It is reputedly beneficial for the lungs. Topaz is the gemstone of the star sign Scorpio.

turquoise

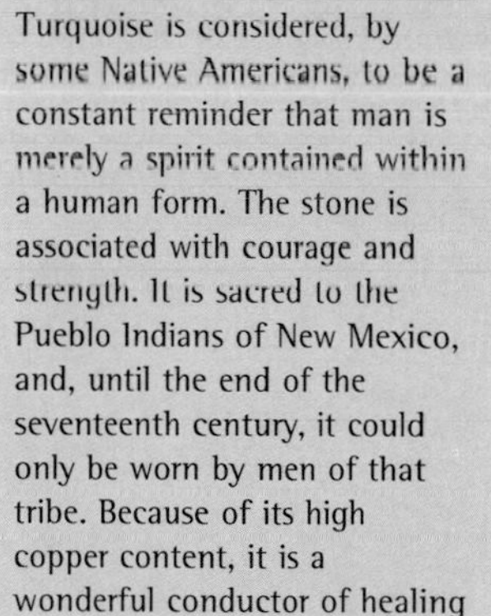

Turquoise is considered, by some Native Americans, to be a constant reminder that man is merely a spirit contained within a human form. The stone is associated with courage and strength. It is sacred to the Pueblo Indians of New Mexico, and, until the end of the seventeenth century, it could only be worn by men of that tribe. Because of its high copper content, it is a wonderful conductor of healing energies.

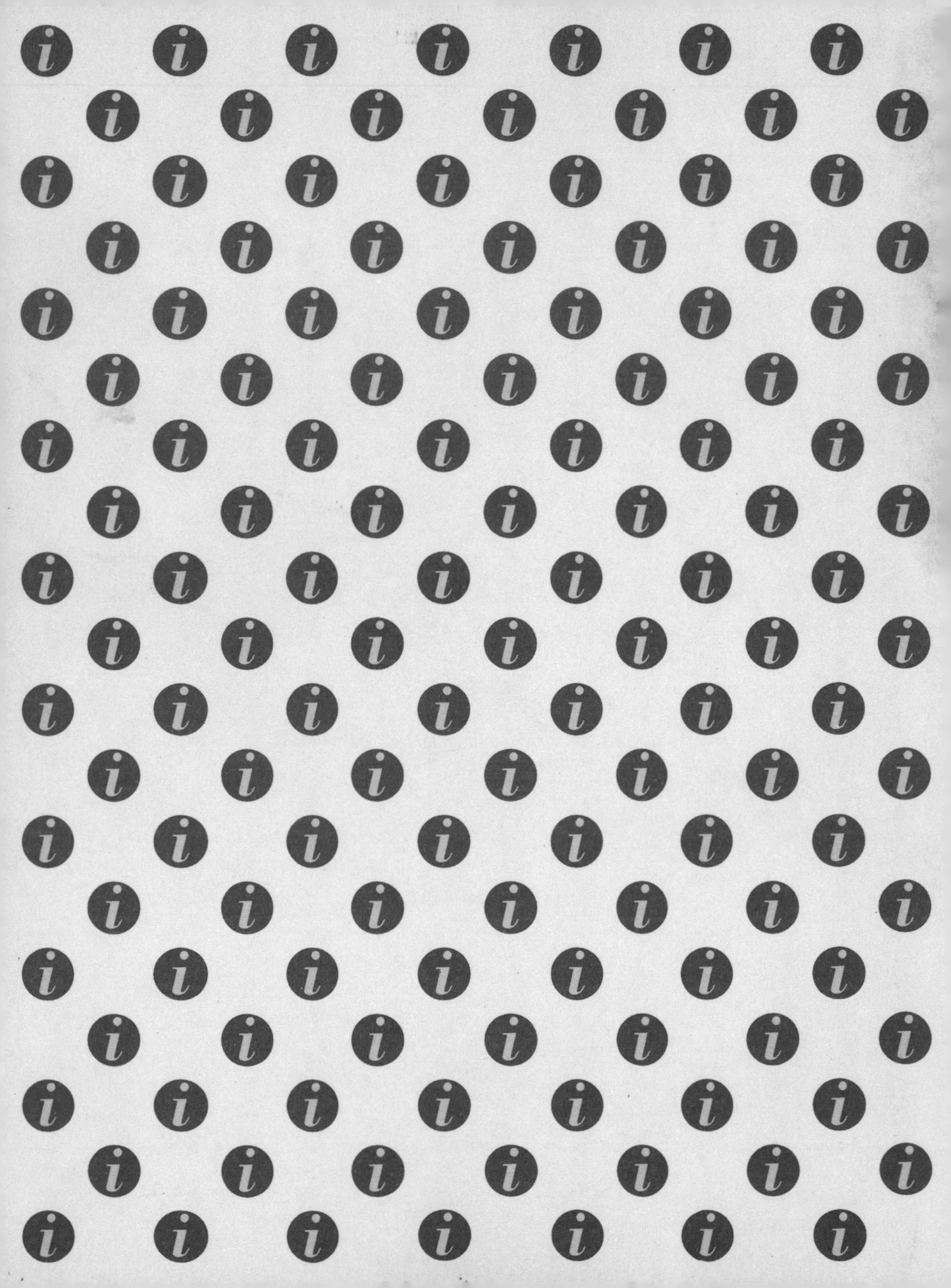